Genetics from Laboratory to Society

Health, Technology and Society

Series Editors: Andrew Webster, University of York, UK and Sally Wyatt, Royal Netherlands Academy of Arts and Sciences, The Netherlands

Titles include:

Andrew Webster (*editor*)
NEW TECHNOLOGIES IN HEALTH CARE
Challenge, Change and Innovation

Gerard de Vries and Klasien Horstmann (*editors*)
GENETICS FROM LABORATORY TO SOCIETY
Societal Learning as an Alternative to Regulation

Forthcoming titles include:

John Abraham and Courtney Davis
CHALLENGING PHARMACEUTICAL REGULATION
Innovation and Public Health in Europe and the United States

Alex Faulkner
MEDICAL DEVICES AND HEALTHCARE INNOVATION

Herbert Gottweis, Brian Salter and Catherine Waldby
THE GLOBAL POLITICS OF HUMAN EMBRYONIC STEM CELL SCIENCE

Jessica Mesman
MEDICAL INOVATION AND UNCERTAINTY IN NEONATOLOGY

Maggie Mort, Tracy Finch, Carl May and Frances Mair
MOBILISING MEDICINE
Information Technology and the Modernisation of Health Care

Steven P. Wainwright and Clare Williams
THE BODY, BIO-MEDICINE AND SOCIETY
Reflections on High-Tech Medicine

Health, Technology and Society
Series Standing Order ISBN 1–4039–9131–6 Hardback
(outside North America only)

You can receive future titles in this series as they are published by placing a standing order. Please contact your bookseller or, in case of difficulty, write to us at the address below with your name and address, the title of the series and the ISBN quoted above.

Customer Services Department, Macmillan Distribution Ltd, Houndmills, Basingstoke, Hampshire RG21 6XS, England

Genetics from Laboratory to Society

Societal Learning as an Alternative to Regulation

Edited by

Gerard de Vries
University of Amsterdam

Klasien Horstman
Technical University Eindhoven, and University of Maastricht

palgrave
macmillan

First published in Dutch in 2004 by Aksant, Amsterdam as
Genetica van laboratorium naar samenleving.
De ongekende praktijk van voorspellende genetische testen.

First published in English 2008 by
PALGRAVE MACMILLAN
Houndmills, Basingstoke, Hampshire RG21 6XS and
175 Fifth Avenue, New York, N.Y. 10010
Companies and representatives throughout the world

PALGRAVE MACMILLAN is the global academic imprint of the Palgrave
Macmillan division of St. Martin's Press, LLC and of Palgrave Macmillan
Ltd. Macmillan® is a registered trademark in the United States, United
Kingdom and other countries. Palgrave is a registered trademark in the
European Union and other countries.

ISBN-13: 978–0–230–00535–8   hardback
ISBN-10: 0–230–00535–7   hardback

A catalogue record for this book is available from the British Library.

A catalog record for this book is available from the Library of Congress.

10   9   8   7   6   5   4   3   2   1
17   16   15   14   13   12   11   10   09   08

Printed and bound in Great Britain by
CPI Antony Rowe, Chippenham and Eastbourne

Contents

Abbreviations

ABI	Association of British Insurers
AMC	Academic Medical Centre (Amsterdam) (Academisch Medisch Centrum Amsterdam)
ARBO(WET)	Dutch Occupational Health and Safety Act (Arbeidsomstandigheden Wet)
BPV&W	Grassroot Organization for Insurance and Labour (Breed Platvorm Verzekerden en Werk)
BW	Civil Law (Burgerlijk Wetboek)
BRCA (1/2)	Breast Cancer Gene
CBO	Dutch Institute for Health Care (Centraal Begeleidingsorgaan voor Intercollegiale Toetsing)
COT	Independant Advice, Inspection and Management for Building and Industry (Centrum voor Onderzoek en Technisch Advies)
CF	Cystic fybrosis (hereditary life threatening disease)
CTE	Chronic toxic encephalopathy (see OPS)
CVZ	Netherlands Health Care Insurance Board (College voor Zorgverzekeringen)
CHZ	Coronary Heart Disease
DNA	Deoxyribonucleic acid
EHC	Hereditary Hypercholesterolaemia Foundation (Stichting Erfelijke Hypercholesterolemie)
FH	Familial Hypercholesterolaemia
FISH	Fluorescente In Situ Hybridization
FME-CWM	Trade organization for the technological-industrial sector (Federatie Metaal- en Elektrotechnische Industrie – Contactgroep van Werkgevers in de Metaalindustrie)
FNV	Dutch Confederation of Trade Unions (Federatie Nederlandse Vakbeweging)
FWO	Research Foundation – Flanders (Belgium)
HDL	High-density lipoprotein (the 'good' cholesterol)
HGAC	Human Genetics Advisory Commission
HRM	Human resources management
ICSI	Intra cytoplasmic sperm injection
i-FISH	interface Fluorescente in situ hybridization
IVF	In-vitro fertilization

JAMA	*Journal of the American Medical Association*
KNMG	Royal Dutch Medical Association (Koninklijke Nederlandsche Maatschappij tot Bevordering der Geneeskunst)
KWF	Dutch Cancer Society (Koningin Wilhelmina Fonds)
LDL	Low-density lipoprotein ('bad cholesterol')
LPL	Lipoprotein lipase ('very bad cholesterol')
LVO	Belgian Law on Insurance Contracts (Landverzekeringsovereenkomst)
MAC	Maximally Acceptable Concentration
NHG	Dutch College of General Practitioners (Nederlands Huisartsen Genootschap)
NIGZ	Netherlands Institute for Health Promotion and Disease Prevention (Nationaal Instituut voor Gezondheidsbevordering en Ziektepreventie)
NIZW	Dutch Institute for Care and Wellbeing (Nederlands Instituut voor Zorg en Welzijn)
NTVG	Dutch Journal of Medicine (Nederlands Tijdschrift voor Geneeskunde)
NWO	Netherlands Organization for Scientific Research (Nederlandse Organisatie voor Wetenschappelijk Onderzoek)
OPS	Organic Psycho-Syndrome (brain disease, often called 'painters disease' or 'Scandinavian disease'; see CTE)
OTA	Office of Technology Assessment
PKU	Phenylketonuria (inherited metabolic disease)
RMO	Netherlands Council for Social Development (Raad voor Maatschappelijke Ontwikkeling)
RNA	Ribonucleic acid
RSI	Repetitive Strain Injury
SER	The Social and Economic Council of The Netherlands (Sociaal-Economische Raad)
STG	The Foundation for Future Health Scenarios (Stichting Toekomstscenario's Gezondheidszorg)
StOEH	Hereditary Hypercholesterolaemia Detection Foundation (Stichting Opsporing Erfelijke Hypercholesterolemie)
StOET	Hereditary Cancer Detection Foundation (Stichting voor Opsporing Erfelijke Tumoren)
TNO	Dutch Organization for Technological Research and Innovation (Nederlandse Organisatie voor Toegepast Natuurwetenschappelijk Onderzoek)

WBO	Organization of Business and Industry Act (Wet op de Bedrijfsorganisatie)
WGBO	Dutch Law on Medical Treatment (Wet op de Geneeskundige Behandelingsovereenkomst)
WHO	World Health Organization
WMK	Dutch Medical Examination Act (Wet op de Medische Keuringen)
ZON	Netherlands Organization for Health Research and Development (Zorg Onderzoek Nederland)

Preface

On 26 June 2000, at press conferences in Washington and London, it was announced that the Human Genome Project and its privately funded competitor Celera had managed to decode the nearly complete sequence of the human genome, 'the blueprint of human life'. It was rightfully welcomed as a first-rate scientific achievement, and some commentators even drew comparisons with the publication of the work of Galileo and Darwin. That President Clinton and Prime Minister Blair were the ones to make the announcement only serves to underline that also from a public point of view, the feat involved was a remarkable one indeed.

Detailed knowledge in the field of genetics is bound to have far-reaching consequences for medicine. More than in the past, medical science will be able to inform us about our chances of contracting specific diseases in the course of our lives. Increasingly, a *risk-oriented, predictive* medicine is emerging. Where medicine plays an important role in our life and social interactions, we have to deal with the effects of this new development. Insights in the area of genetics will have implications not only in diagnostics, treatment and the prevention of diseases, but also in terms of our lifestyle, insurance, the extent to which we are held responsible for our own health, for family ties, labour relationships and overall social solidarity.

In the years prior to the announcement of 2000, the social ramifications and implications of the rapid developments in the field of genetics were already discussed extensively. A wide variety of authors and institutions have voiced ethical and political concerns about the foreseeable progress in the area of genetics and predictive medicine, as well as about the practices of clinical genetics and prenatal chromosome research as they evolved in the 1980s and 1990s. In 1999 the editors of the present study, in collaboration with O. Haveman, published *Gezondheidspolitiek in een risicocultuur – Burgerschap in het tijdperk van de voorspellende geneeskunde* (Health politics in a culture of risk – Citizenship in the age of predictive medicine). The publications have triggered discussions on TV and in the press, in forums, political parties, the medical world and in the insurance industry.

The rapid developments in the field of genetics do not only urge us to deal with the *potential effects* of genetics and how society ought to respond; they also challenge us to address the issue of *how to discuss* the concerns involved in a meaningful way and how, as a society, we should cope with the presumably far-reaching yet largely unknown possibilities of this new technology. Should we design meticulous regulations to channel the developments involved and curtail the undesirable social effects? If the related technical developments are decidedly international in scope, is it still useful to have national governments formulate national regulations? Will a DNA-test that is prohibited in one country not be available through the Internet? Once sociologists, ethicists, and politicians are ready to address these sorts of concerns, haven't they been basically settled already by researchers and clinicians? Do the developments in genetics not simply outpace our ability to control them by legislation, leaving it up to parliament merely to legitimize recently established practices a posteriori?

Over the past decade, detailed knowledge on the human genome has become available. However, we still know little about the practices in which such knowledge will be applied. Nor do we have much insight into the nature of the various social consequences of this new technology. On these issues, even experts who are closely involved are fumbling in the dark. They, too, have great trouble assessing the new scientific developments in terms of their social implications. If the human genome is a highly complex system, the same is true of society.

The present collection describes and discusses various aspects of different practices in which genetic knowledge and technology play a role, like health care, insurance and the labour market. It does so by emphasizing that genetic technology – instead of providing new certainties that determine our lives and society more and more – introduces many uncertainties. Moreover, it will become clear that in making this technology 'work' new responsibilities emerge and old ones are redistributed. In other words, the study shows that the notion of 'implementation' is far too simple to catch the work that is asked from professionals, citizens, regulative agencies, insurers, employers and employees in the trajectory from laboratory to society. Therefore social, ethical and political problems are more diverse than the ongoing debates on genetics suggest. This being the case, the familiar ways of tackling these problems – through public regulation, professional standardization and individual autonomous decision-making – are not necessarily the most appropriate. Together with the rise of genetics, traditional relations between the state, professionals and citizens are being transformed and we cannot assume that established relations – for example, between doctors and their patients, or between insurance companies and their clients – remain unaffected. To do justice to these developments in technology as well as in society, this study argues for discussing the trajectory of genetic technology from laboratory to society not in terms of implementation, but in terms of sensible social experimenting. We have to experiment and learn about genetics and society at one and the same time, and the different contributions will provide input to that learning process.

The chapters in this book are based on research carried out at the University of Amsterdam, the University of Maastricht and the Catholic University of Leuven. This research was made possible in part by a grant from the NWO Programme on Social Cohesion – Social Participation, Commitment and Involvement, and by subsidies from the Netherlands Organization of Health Research (ZON) and the Belgian FWO. The authors would like to thank these funding providers for their support. This book was published in Dutch in 2004 (*Genetica van laboratorium naar samenleving. De ongekende praktijk van voorspellende genetische testen*, Amsterdam: Aksant). We gratefully acknowledge the work and comments of Ton Brouwers, our translator, and we thank NWO for providing the support that made this translation possible.

Amsterdam/Maastricht
November 2006

Notes on the Contributors

Ruth Benschop is a post-doctoral researcher at the Faculty of Arts and Social Sciences, University of Maastricht, The Netherlands.

Marianne Boenink is Assistant Professor of Philosophy and Ethics of Biomedical Technology, Department of Philosophy, Faculty of Behavioural Sciences, University of Twente, The Netherlands.

Klasien Horstman is Socrates Professor of Philosophy and Ethics of Bioengineering at the Technisch University Eindhoven, and Associate Professor at the Faculty of Health, Medicine and Life Sciences, University of Maastricht, The Netherlands.

Ine van Hoyweghen is a post-doctoral researcher at the Faculty of Health, Medicine and Life Sciences, University of Maastricht, The Netherlands.

Carin Smand studied Health Sciences at the University of Maastricht and is Coordinator-Manager of the Dutch Transplantation Foundation.

Rein Vos is Professor of Health Ethics and Philosophy at the Faculty of Health, Medicine and Life Sciences, University of Maastricht, The Netherlands.

Gerard de Vries is Professor of Philosophy of Science at the University of Amsterdam, The Netherlands, and Council Member of the Scientific Council for Government Policy in The Hague, The Netherlands.

Myra C. B. van Zwieten is Lecturer in Medical Ethics and Communication at the Academic Medical Centre, Clinical Methods and Public Health Division, Department of General Practice, Amsterdam, The Netherlands.

1
The 'Unknown' Practice of Genetic Testing

Gerard de Vries

The rise of predictive medicine

In the twenty-first century physicians will devote an increasing part of their time to long-term health risks rather than to acute diseases. Of course, doctors will continue to listen to the health complaints of their patients and, if necessary, they will provide medical treatment. But the emphasis in medicine will shift from crisis-driven intervention to prediction of individual health risks and to the question how such risks can be managed or reduced. Aside from the complaints-bound clinical medicine everyone has become familiar with in the nineteenth and twentieth centuries, *predictive medicine* has emerged. Over the next decade or two, it is likely that entire populations or specific subgroups will be screened for genetic information in order to target interventions on individual patients that will improve their health and prevent disease.

The rise of predictive, risk-oriented medicine will have important ramifications for the population at large. Not only physicians, but all of us will increasingly pay attention to health risks and to the measures we should take to prevent potential illnesses and handicaps from becoming realities. We will take risk assessments into account when discussing our way of life, choose a diet with health risks in mind, start thinking about pregnancy in terms of a child's chance of having a handicap, and begin to look at the genetic make-up we in part share with our relatives as a source of future concern. Governments and other well-meaning organizations that already expose us to all sorts of advice on how to live healthy lives will increasingly do so by stimulating us to take our individual risk profile into account when choosing our lifestyle. In the future, people with a genetic disposition that puts them at increased risk of cardiovascular diseases or rheumatism may get their groceries from a special section in the neighbourhood supermarket. And if *we* are not yet interested in our individual health risks, our employers and insurers probably are.

In contrast to older forms of preventive medicine that addressed large sections of the population, modern risk-oriented medicine targets *individual* health risks. The warnings that smoking, eating too much fatty food and not exercising enough can cause cardiovascular diseases have by now resounded for several decades. Based on epidemiological studies, this type of public health information is a form of predictive medicine. There is broad consensus, however, that the importance of predictive medicine will rise strongly due to the developments in genetics over recent decades. In cases where the individual risk for a disease can be determined with a high level of reliability, it becomes possible to take measures based on that knowledge to prevent disorders at the individual-specific level. While previously public health information had a rather general character, advice will now become much more detailed and more personalized.

Risk-oriented, predictive medicine will give rise to a new kind of health-care practice and eventually to another role for medicine in society. It may lead to unprecedented medicalization of everyday life. It will influence our way of thinking about health and disease. This is already the case in professional medical thinking. The clinical medical science that we are all familiar with is based on a concept of disease that defines it (according to a much-cited definition) as 'an internal state which is either an impairment of normal functional ability ... or a limitation on functional ability caused by environmental agents' (Boorse cited in Temple, McLeod et al. 2001). Those who are ill will notice this from the – partial – dysfunction of their body or mind; they have specific complaints. To take this new predictive medicine into account, a new definition has recently been proposed, according to which a disease is 'a state that places individuals at increased *risk* of adverse consequences' (Temple, McLeod et al. 2001) On this account someone may suffer from a disease when there are neither acute complaints nor any other signs of dysfunction. Under predictive medicine, 'disease' refers no longer only to the present state of one's body or mind, but also to its possible condition in the – perhaps distant – future. The scope of the concept of 'disease' is significantly extended.

Shifts in thinking about disease like this are not just academic niceties. They may have profound social consequences. Due to the development of predictive medicine, people will increasingly be held responsible for their health. After all, if a disorder can be predicted and options to substantially reduce the likelihood of its occurrence are available, the disease is no longer an inevitable fate that just happens to people. Those who were aware of the risks and yet fell ill may easily be considered as persons who took insufficient care of themselves; who failed to manage their own lives properly. Rather than being considered a victim of a disease, they become perceived as the perpetrator or as complicit. This shift in thinking about disease will likely put pressure on solidarity and on the foundations of socially organized health care and the insurance system (De Vries 2003).

Genetics-based predictive medicine is not only a matter of the future. DNA diagnostics, chromosome testing and heredity counselling have already become established practices in the 1980s and 1990s. There are by now a number of tests available that can predict with a high degree of reliability whether someone has a significantly increased risk of contracting a serious disease during their lifetime. These include tests for Huntington's disease, cystic fibrosis, Duchenne's disease, hereditary hypercholesterolemia, as well as for some forms of breast cancer and colon and prostate cancer. Thus far, genetic testing mostly involves tests that can map defects in a single gene or a very limited number of genes. Nearly every day, however, new genetic disorders are discovered that can be linked to a particular disease. Many experts expect that at some point predispositions to the more common diseases will, in the same way, also be detected early and that there will be tests for allergies, occupational diseases and various widely prevalent diseases (such as frequently occurring cancers, cardiovascular diseases and Alzheimer's disease). Genetic screening (testing offered systematically to specific groups – for instance, all pregnant women, or all newborns) and genetic detection (targeted testing, such as within one family) are currently performed for some disorders. Prenatal diagnostic tests for Down's syndrome that provide near certainty are offered in many countries to pregnant women. Nationwide detection of hereditary hypercholesterolemia is in place in some countries, while the implementation of a dozen other programmes for genetic screening or detection is under discussion.

When it is found that someone is genetically predisposed to a particular disease, that person may subsequently decide to change their diet or radically adapt their lifestyle, either to lower the chance of the disease's manifestation or to reduce its effects. In other cases people will be advised to take drugs and to enter a programme to regularly check their condition. In some cases preventive surgery may reduce specific risks. When a prenatal test establishes a high probability that the foetus is suffering from a serious disorder, it may be decided to discontinue the pregnancy. Even in cases where no therapy is available, as in Huntington's disease, a test can still be helpful, if only to end agonizing uncertainty.

It is not only diagnostics that will benefit from the advances made in genetics. In the area of treatment, too, new developments are to be expected. The pharmaceutical industry invests heavily in the development of drugs that are effective in the treatment of patients with a specific genetic profile. As a result, it is anticipated that the practice of having to administer various medications on a trial and error basis merely to select an efficacious treatment for a particular patient will gradually become a thing of the past. Much work is also being done on functional foods that can help people with a specific genetic profile to reduce their risks.

Public debates about genetics

The development of risk-oriented medicine and the expected role of genetic testing in particular have triggered widely divergent reactions. From the 1990s on, genetics and its consequences for medicine have become issues of public debates (cf., for instance, Buchanan, Brock et al. 2000; Habermas 2001; Horstman, De Vries and Haveman 1999; Jasanoff 2005; Kitcher 1996; Tijmstra 2004). In these debates contributors exchanged their hopes and fears. For many the development of risk-oriented predictive medicine holds the promise of realizing an old dream, namely that of a medical science and a health-care system that aim to prevent the occurrence of disease as much as possible, instead of intervening only after the symptoms of any particular disease are manifested. However, aside from this optimism, concerns have also been aired. It is widely acknowledged that the developments in genetics may confront individuals and society with serious ethical and social problems. Individuals will have to face often painful dilemmas. How should one deal with the knowledge that one has a genetic predisposition for contracting a serious disorder in the long term? Should surgical removal of both breasts be recommended to a young woman who learns she carries a mutation of the BRCA1 or 2 gene, meaning that she has an increased chance of developing breast cancer within the next two or three decades? How should those involved weigh the statistical information on things that might happen at some point in the future against the certainty of a very unpleasant intervention? And how are family relationships influenced when information about a genetic disease becomes available that affects some members of a family but not all?

The problems are not just confined to the private sphere. People who have undergone diagnostic genetic testing may run up against barriers once they need to purchase life insurance, for example, for a mortgage. Should a physician in fact urge patients to take a genetic test if he or she knows that they are likely to encounter this kind of problem? Or should perhaps the right to get insurance – up to a certain amount of money – be guaranteed by law to balance the foreseeable social disadvantages of genetic testing? And what are the consequences for the rights of employees when employers, before selecting staff, could require applicants to take genetic tests for specific occupational diseases? Does the application of genetics in labour contexts contribute to creating new forms of discrimination and inequality, and perhaps even to the emergence of a 'genetic underclass'?

Critics have also advanced concerns of a more fundamental nature. It has been claimed that the deployment of genetic techniques would entail the spectre of eugenics in a new guise. The German philosopher Jürgen Habermas has argued that the future of human nature is at stake when it comes to specific genetic techniques: what so far had to be accepted as given, now seems to have become a matter of choices and options. How does this affect our view of human life? Do the genetic technologies currently under development irrationally exacerbate the apprehension that life comes with its own misfortunes? What does it tell us about society if children born with foreseeable deficiencies are perceived as the results of parents failing to take responsibility, or as a health-care expense that could and should have been avoided? And are these grave social problems really an inevitable price for our desire to improve our health? It is already possible, after all, to gain major health benefits by opting for a sensible lifestyle on the basis of the existing knowledge. So shouldn't we keep genetics-informed predictive medicine at arm's length for the time being and first take the available measures that would promote our health: stop smoking, improve our dietary habits and start exercising – not to mention heavily invest in health care in poor countries?

The rise of genetic testing confronts society with serious problems, as both sceptics and proponents of the introduction of genetic technologies in medicine concede. In public discussions, these problems are usually presented in the form of dilemmas. How should the balance between benefits and social dangers be achieved? How, in concrete cases, should one decide about the use of a test and its results? There is a broadly felt need for standards, and for legislation and regulation to ensure the proper course of the various developments involved. Aside from technical aspects, which are generally conceived as being of prime concern for medical experts, the issues clearly have major ethical and social dimensions. From the 1980s on, debates have been conducted in all industrial nations, sometimes on genetics in general, in other cases with a more specific focus. Several conferences have explored the ethical and social issues at stake. The combination of the knowledge of experts and the perspectives of laypeople from various religious and political backgrounds was thought to provide a sound basis for social consensus and effective regulation.

Now, after more than two decades of debates on genetics, it should be concluded, however, that many of the early public discussions were based on quite optimistic, if not exaggerated, expectations about the potential of future developments. Both advocates and opponents of genetic diagnostics have been guilty of this. Eager to get funding for their work, researchers in genetics tended to overstate the potential revenues of their work, while underestimating the time needed for developing practical, clinical applications. What better reason to spend taxpayers' money than the promise that it will contribute to finding a cure for Alzheimer's disease, Parkinson's disease or cancer? Not surprisingly, biotech and pharmaceutical companies that invest substantial amounts of money in genetic technology joined in. Features of future products that in some cases had not even left the research laboratory were broadly advertised. Interestingly, in many cases those who underscored the ethical and social problems of genetic technologies simply accepted these promises at face value, only to emphasize the darker sides of the projected developments. As a result debates were frequently based on untenable claims and quite regularly on the overly optimistic view that new technologies work out in the ways their designers have in mind. Even a cursory overview of the history and sociology of technology, however, shows that reality is more unruly than that. Typically, the road from basic research to practical applications is long and winding and the journey often leads to unanticipated destinations. When assessing the implications of new technologies, we have to keep in mind that the phrase 'works as advertised' is first and foremost a commercial slogan.

Biological and epidemiological considerations impose serious restrictions on the possibilities of testing. It is still the case that little is known about the relationship between genotype and phenotype, the ways in which various genetic factors influence each other and the interactions between genes and environmental factors. Genetic determinism has no basis in science. In fact, prominent experts, including the leader of the British division of the Human Genome Project, John Sulston, have expressed doubts on whether the complex relationships involved will ever become sufficiently known to actualize the projected practical application of genetic techniques on a large scale (Sulston and Ferry 2002, ch. 7). The first, striking successes of clinical genetics may easily mislead us. They were achieved mainly with rare, monogenetic diseases that occur in families with a known history of the disease. It is incorrect to think that screening of the entire population or of large groups for multi-genetic diseases merely involves extrapolation of the current methods in clinical genetics. Methodological problems, the laws of statistics and practical limitations stand in the way of rapid successes.

Eventually, the impact of the broad public discussions of the 1980s and early 1990s turned out to be quite limited. Gradually, the moral issues involved have been moved outside the public domain to become matters of private choice. From an exchange on the public effects of the introduction of genetics in quite general terms, the debate has shifted to the question whether specific tests ought to be permitted and which concrete measures are called for in order to address the anticipated ethical concerns. The discussion has evolved into a format in which genetic techniques are first evaluated exclusively for their medical merits. Tests that, in the eyes of medical professionals, have clear-cut medical advantages are introduced. A combination of professional monitoring and the guarantee that individual patients will always be free to choose whether or not to make use of the tests offered is supposed to provide sufficient warranty to cover the ethical and social issues. With genetic techniques that have no direct medical application, or that as of yet are deemed unreliable, more restraint is exercised.

As a result of this newly emerged format, the broader social and political considerations that were discussed in the early stages of the debate still play only a modest role at best. As far as technical issues are concerned, trust is put in the established procedures for self-regulation in the medical world. Ethical issues are turned into a private matter, thus effectively removing them from the public forum. Except for population screening (for which official government approval is required in most countries), the role of the public at large, their political representatives and government authorities in determining the future of genetic testing has indeed become quite limited.

Piecemeal approach

With the emergence of the new format around 2000, the debate on predictive medicine and the role of genetics has gradually taken on a different character. Today, the debate is conducted primarily by the medical professionals who will directly work with the new techniques and by ethicists and health law experts specialized in this subject. A *piecemeal approach* has become dominant: the problems that emerge are examined separately, on a step-by-step basis. Issues pertaining to the putative medical value of a test are thereby distinguished from its projected ethical problems. The former are primarily decided by the medical professionals involved and in some cases by insurance companies, namely when the question arises whether a test should be part of their services package. The ethical questions are answered in terms of legislation and regulation.

The solution that time and again presents itself is one whereby the application of a new technique is put under professional monitoring with eligible clients subsequently being left to decide individually whether they want to use the available possibilities or not.

The question to whom and when tests are offered, if at all, is thus primarily seen as a matter for the medical professional world. Whether or not a person has any tests that are offered, in fact permits their performance, is subsequently conceived of as a question that the individual must answer. He or she has to face a difficult choice with both medical and ethical aspects as well as potentially far-reaching consequences. Medical experts will of course provide guidance and they will first of all make sure that the person is well informed about the various options, the major technical aspects and the potential consequences of a specific choice. If anything, the decision needs to be made on a solid factual basis. Regarding the ethical aspects, however, the medical expert is expected to show much restraint. Ultimately, whether or not someone will decide to undergo a test should always be a matter of free, so-called 'autonomous' choice.

The piecemeal approach relies on ethical principles that guide clinical practice and that are broadly shared and in many countries are backed by law. The ideal of the physician–patient relationship and the principles of 'patient autonomy' and 'informed consent' that have been adopted in the second half of the twentieth century in clinical medicine have been exported to predictive medicine. This should not come as a surprise as the major part of predictive medicine is practised within institutions in which clinical care is also supplied, or in institutions that are closely tied to it. With paternalistic relationships between physicians and patients in clinical practice belonging to the past, few would welcome a reintroduction of such relationships in medical practices in which genetic tests play a role. In fact, in clinical genetics the protection of the autonomy of the patient is emphasized even more stringently than in the longer-established forms of medicine (Nelis 1998).

The piecemeal approach's major advantage is that discussions on the use of medical techniques become more realistic, while attention gets centred on tests that are genuinely applicable in actual practice. Speculation and discussions based on science fiction scenarios are thus avoided. This approach also has its drawbacks, though. It addresses the rapid development of genetic techniques as a fact to which society and the community of medical professionals has to respond. The response is geared towards the challenges that concrete tests present. Existing professional relationships and forms of self-regulation, together with ethical advising, are assumed to provide a sound framework for responsible judgments on benefits and disadvantages. However, by concentrating on concrete, directly manageable problems, long-term and secondary effects easily disappear from view. The logic of the step-by-step approach is that of focusing all attention on the points known to be in clear danger – in the case of a threat of flooding, sandbags are brought in and, where needed, one builds a dam. But the question whether these activities may cause the river gradually to change its course and bed and bring about new problems elsewhere remains unconsidered.

Secondary effects and the question to what degree step-by-step approval of genetic techniques will lead to institutional changes and new relationships deserve, however, to be publicly discussed. Genetic testing requires the input of representatives of disciplines that in traditional clinical medicine are thought not to be directly involved in clinical medical work, such as epidemiologists, statisticians and communications experts. Moreover, it is often necessary to call upon institutions such as databanks and population registers, causing new issues – for example, matters of privacy – to arise. Due to the role of disciplines traditionally not conceived as medical specialties, as well as of non-medical institutions, boundaries are blurred and the existing system of quality control for health care is put under stress. Whereas professional bodies and legal authorities carefully watch the professional activities of physicians, the same is not yet true for the communication experts who help to write the leaflets informing future patients about what they may expect when they ask for genetic advice. Moreover, genetic testing may require the involvement of non-professionals. In some cases a test's success will even depend on the knowledge people have of their family, and on their persuasiveness and the effort they put in to have their relatives tested as well. The effects of the new roles of such professional groups, institutions and laypeople on the possibilities and limitations of self-regulation, on which the piecemeal approach to a large extent relies, deserve attention. Cultural effects should also be a matter of concern. The combination of clinical-genetic practice, public health education, and occupational and insurance medicine – all relying on the same categories and distinctions of risks – may easily produce a situation in which practices that each started out as sensible and geared to an individuals' wellbeing together evolve into a cultural straightjacket, with disease and disability appearing as unsuccessful attempts at taking one's life into one's own hands. New forms of discrimination and exclusion then lie ahead.

The ease with which the autonomy principle that has proved its merits in clinical medicine is adopted in predictive medicine deserves critical attention. This move is far from obvious. Why has it occurred without much ado? There is little reason to suppose that the population as a whole adheres to the liberalist ethics that has autonomy as its cornerstone. Of course, many will accept the autonomy principle based on their conviction. But in pluralist societies, we cannot expect ethical liberalism to constitute the generally accepted ideal. Liberalist ethics, however, has a clear-cut advantage: it allows for legal regulation while acknowledging differences in ideology by reserving ultimate choice to the individual. As such, it offers itself as a form of governmental rationality particularly suited to pluralist societies. It provides a clear-cut, practical, political solution in situations where general consensus about ethical issues may not be expected and the need for regulation on ethical matters is nevertheless conceived as pressing. By accepting the principle of autonomy, difficult ethical questions are removed from the public domain and relegated to the private sphere. Of course, those who subscribe to a liberalist ethics will welcome this move, convinced as they are that the ethical issues arising in modern medicine should have been considered as private matters from the start. The effect of this approach, however, is that normative problems disappear behind closed doors: they are confined to discussions among medical professionals, to interactions between care providers and their clients in the privacy of consultation rooms, and to the private thoughts of patients and discussions in their families. Even if we assume that these discussions take place conscientiously – which in most countries we have no reason to doubt – a major element is absent: little is learned from them at the *public* level. Perceived from a *public* point of view, to prematurely, without serious discussion, adopt autonomy as the guiding ethical principle in predictive medicine is little more than the ostrich policy of burying one's head in the sand.

'Unknown' practices and their public issues

In his remarkable book on politics, *The Public and its Problems*, the American pragmatist philosopher John Dewey defined 'the public' as consisting 'of all those who are affected by the indirect consequences of a transaction to such an extent that it is deemed necessary to have those consequences systematically cared for' (Dewey 1927, pp. 15–6). The public, then, is not a *given* body, but an assembly of people relative to an issue. They may, or may not, know each other and they may not even be aware of each other's existence and of the fact that they share interests. An important function of the democratic state, according to Dewey, is to *make things public* – that is, to help organize the public, to bring the adverse consequences of the actions they were not involved in into view, and to subsequently deal with the adverse consequences; for example, by declaring that the activities from which they originate are outside the law, or by providing adequate compensation for those who involuntarily have to face these consequences.

For a wide range of issues, states have both the means and the institutions to perform this function. Today, for example, all modern states have agencies that monitor pollution, identify those who suffer from it, and compensate them for damage or punish industries that are responsible for environmental damage. In other cases, however, societies still have to *learn* to what extent the adverse consequences of private transactions should be 'deemed necessary to be systematically cared for'; they have to learn which public is involved or at risk of becoming involved, how it may be organized and how the adverse consequences at stake are to be dealt with. In these cases, in which it is still largely unknown to what extent a *public* issue is at stake, uncertainty reigns and the issues that emerge may be called 'unidentified political objects' (Dijstelbloem, Schuyt and De Vries 2004). It often requires a lot of time and the efforts of many – both within and outside traditional political arenas – before the issue at hand and the public involved are identified with sufficient precision for concrete policy proposals to be fruitfully adopted. How is the public to be organized? What are the appropriate arenas for discussing the consequences of a practice, and which official bodies need to take care of handling the adverse consequences? These are questions that in these cases still need to be resolved.

Industrial societies are increasingly faced with 'unidentified political objects', especially in relation to new technologies (Beck 1992). Genetic testing is clearly one of them. In many respects it is still far from clear who is involved as an affected public, to what extent genetic testing produces consequences that should be systematically addressed, or how – as well as by whom and where – reasonable policies to cope with this issue can be deployed.

The uncertainties we have to deal with in this domain are of at least two kinds. In the first instance, as outlined above, genetic testing is still largely a matter of the future. Available tests are of recent date and pertain to a limited number, mostly quite rare, diseases. It is doubtful to what extent the current experiences with clinical genetics that apply to the diagnostics of fairly rare diseases in families with known case histories can be extrapolated to future forms of genetic detection and screening that will involve more common, multi-genetic diseases. Although knowledge of genetics has significantly advanced over recent decades, it should also be acknowledged that only limited experience has been gained with the *practice* of genetic testing. Which issues will arise and which publics will be affected depends on a wide array of factors, both technical and scientific, but also on how genetic practices become organized and socially embedded, as well as on the future conditions for insurance. No one can claim to have expert knowledge on such matters. Precisely because many contingencies interfere, prediction on this score is pure speculation. On questions pertaining to the future social role of genetics, all of us, including the best experts in genetics, fumble in the dark.

Genetic testing is also 'unknown' in another sense. Even if we turn our attention to already *existing* practices, there is still a lot to be illuminated. Of course, the medical experts who work in those practices can inform us about what is going on, and individuals who have undergone genetic testing may inform us about their experiences. However, it is often the case that those who are closely involved in a practice fail to notice much of what is going on in it, simply because they take many things for granted. They may fail to perceive relevant aspects of the practice, not because they are ignorant or because these aspects are covered-up, but because they are too close, too immediately linked to them.

A simple, well-known example may demonstrate this second sense in which we may speak of a practice as being 'unknown'. When asked for the average speed of automobile traffic, we will have our answer ready available. Some 60 kilometres per hour we say, thereby taking into account that our journey will involve driving on motorways, some unavoidable traffic jams and slow urban traffic. However, in what has become a famous argument Ivan Illich has shown that, in spite of our daily experience with motorized traffic, this estimate is quite erroneous (Illich 1974). To be able to drive that journey, Illich pointed out, we had to first put in a lot of work and time that we simply failed to take into account when giving our answer. To get a driver's licence we had to take many hours of lessons. We have paid road taxes that are used to build and maintain the roads on which we travel. To pay for these taxes we have had to spend quite some time at work to earn money. To pay for our car, for maintenance and gas we had to put in another large number of hours. And so on and so forth. On average we drive some 15,000 kilometres a year. How much time goes into that? If we calculate the hours invested before we can even start the engine, the average speed quickly drops below the 60 kilometres per hour we were thinking of. Making a rough estimate, Illich arrived at an average of some eight kilometres per hour – the speed of someone walking at a stiff pace.

The error that Illich exposed is widespread. When we speak about technologies, we often react like a Russian farmer Tolstoy once introduced.

When he first saw a train, this farmer instantly knew that the train was propelled by its plume of smoke. The very first thing that had struck this man, Tolstoy writes, he had then perceived as the train's driving force. Similarly, we think we have a good notion of what a technology entails when we have described the very first things we notice, such as the device itself or the person who operates or uses it. The infrastructure that has to be in place, the training that precedes its operation, the maintenance needed, the regulations and the monitoring required for its sound and safe functioning – all of this usually remains outside the scope of our considerations. We simply take all of that for granted. By grasping technology merely as involving a ready-made piece of equipment and its users, much of the effort that is crucial for its actual functioning is rendered invisible.

When discussing genetic testing, we easily fall into the same trap. We talk about 'tests', about those who undergo them and about the physicians who are in charge of conducting the tests – these, after all, are the most striking aspects. The normative questions we raise are basically grafted onto the situation thus framed. Should a particular test be introduced? If so, how should we regulate the relationships between physicians and those who wish to be tested? However, in this case too, much of the work that has to be done before testing becomes possible at all is rendered invisible and we forget to raise normative questions about this invisible work. Even the language in which we frame our concerns misleads us. To speak of 'testing' individuals wrongly suggests that certainty about the future health condition of those who undergo a test will be provided within a short time span and in a transparent way. This, however, is seldom the case. 'Testing' comes with new sorts of uncertainties and producing a dependable result will require much time and effort on the part of both medical professionals and the people tested. In fact, those seeking certainty about their genetic predisposition will have to live for extended periods of time with a lot of uncertainty.

The two kinds of uncertainty we have distinguished call for different reactions. When it comes to the future course of genetic testing, all we can do today is speculate. The least that may be expected from us, then, is that we openly acknowledge that we are building castles in the sky. Scenarios about the future should be approached with a sustained scepticism. In many cases, they tell us more about the state of mind of the scholars who made the projections than about the developments that will eventually materialize (De Wilde 2000). This, however, does not imply that we should not be concerned about the future. Acknowledging the uncertainties that are involved, another, more appropriate, attitude is called for. Rather than discussing speculative projections, we should acknowledge that society will have to *learn* how to deal with the various aspects of the new technology. Once we have acknowledged this, we can start to discuss how the learning process can best be organized to promote the likelihood of a reasonable outcome.

As a first step to this learning process, a wide variety of *existing* practices in genetic testing can be studied. To obtain a realistic picture of the public involved and the consequences in terms of the need for systematic care, more information about what is going on is a first prerequisite. When getting on with this task, we will, however, have to face the second kind of uncertainty we identified. After all, not just future practices of genetic testing are 'unknown'; the same applies to the existing ones, not only to society as a whole but also to the experts active in this field and those of us who make use of their work. Much of the work that takes place in genetic testing practices – and hence the uncertainties and responsibilities involved – remains invisible, not because someone has reason to hide anything, but because our way of talking about technology tends to narrow our view. If we are to make explicit what is implicit and taken for granted, practices need to be considered from other angles, not just those of people who either work in them or make use of their services. The detached tools of sociological, anthropological and philosophical study can be utilized for this very purpose.

The present volume is written to contribute to this task. In the course of the book, we visit a wide diversity of sites where genetic testing is currently practised, describing and analysing the work done by the various actors involved. Starting with detailed analyses of what goes on in clinics and laboratories involved in prenatal diagnostics and chromosome testing (Chapter 2) and clinical genetic counselling for hereditary breast cancer (Chapter 3), our focus will gradually shift to other sites and wider circles. This volume includes studies on a programme aimed at detecting those who are at risk of hereditary hypercholesterolemia (Chapters 4 and 5), and on the role of genetic testing in the life insurance industry (Chapter 6) and in occupational medicine (Chapters 7 and 8). The existing practices of genetic testing will be studied from various perspectives, including, but not exclusively so, those of the professionals who work in them and the clients who make use of their services. By following the medical experts and their patients in their daily work, we hope to contribute to making explicit what in most discussions about genetic testing is taken for granted. Before we can raise normative questions about genetic testing in a sustained manner in public, the practice of testing itself in its various dimensions needs to be made public.

The contributions in this volume thus aim to make visible and debatable in public what currently goes on in the private sphere of the interactions between medical professionals and their clients. The chapters in this volume identify specific themes and problems that need to be addressed in open discussions on the future of risk-oriented, predictive medicine. This focus is not so much meant to allow for detailed regulation and a new system of public control, but rather to enrich discussion on the selection of medical techniques by juxtaposing a variety of vantage points. Our closing chapter, Chapter 9, presents a summing up of our results and a discussion of their specific consequences for political and social learning.

Genetic testing involves decision-making by a wide variety of actors, professionals as well as others, and may also have consequences for those not directly involved – the public. Instead of unreflectively accepting that the existing institutions and procedures are adequate to settle the issues involved, this volume seeks ways to *make things public* and to reasonably and democratically decide on the proper course to be taken (Latour and Weibel 2005). The currently dominant piecemeal approach that situates technical issues in medical circles and delegates ethical problems to the 'autonomous' choices of patients is only one of many possible ways of dealing with the public issues in this domain. This volume starts from the premise that to discuss seriously the public issues involved in the practice of genetic testing, society at large first has to learn more about what in fact is going on. The public may be ill-suited to the premature adoption of this piecemeal approach.

This book is written from the view that collective learning requires knowledge about variation. Accordingly, it tries to render such variation visible by articulating how medical professionals and their clients experience genetic testing, and how specific institutions have differed on how to evaluate this practice. The analyses found in this book reveal a large number of uncertainties that are inherent to laboratory work, clinical-genetic diagnostics, usage of genetic tests in medical examinations, technology development and the ways in which innovations lead to institutional changes. The emphasis is on showing plurality, the articulation of uncertainties and differences. Studying existing practices, we zoom in on the work that is actually done and try to make visible the responsibilities that divergent parties have to take on. The descriptions of the various practices involved in genetic testing reveal that choices involving genetic testing are not made by a single party – the medical profession, or the medical profession in dialogue with government. Other parties – such as insurance companies, employers, unions, patients and their relatives – also have their stakes. They make decisions, explicitly or implicitly, and thus take on responsibilities. Our analyses emphasize how such responsibilities are taken and constructed, as well as how developments in genetic testing are linked to shifts in the attribution and distribution of responsibilities.

This book thus distinguishes itself in several respects from the currently dominant approach to the public problems that we have to face in relation to genetic testing. First, the development of genetics is couched explicitly in terms of the rise of a predictive, risk-oriented medicine. Second, attention is devoted to changes in the meaning and use of technology, notably to how technology is accompanied by changes in the ways in which risks are defined and responsibilities shift. Third, we study how genetic information has implications for relationships within the medical-professional world as well as for related sectors (notably insurance and labour relations), whereby social developments that occur outside of health care are taken into account as well. In our closing chapter, finally, normative questions are approached in a different way than those suggested by the usual frames for normative discussions on genetic testing – which are strongly informed by doctor–patient relations. Before we can start discussing detailed matters of regulation, the question of how society can *learn* to deal with genetic testing has to be addressed.

2

Constructing Results in Prenatal Diagnosis: Beyond Technological Testing and Moral Decision-Making

Myra van Zwieten

Grey results

The Roothaan couple are sitting in the waiting room of the prenatal diagnosis outpatient clinic for their appointment with the clinical geneticist. Mrs Roothaan (38) had a chorionic villi sampling a week ago. Yesterday late afternoon she received a telephone call from the hospital that still worries her. Her clinical geneticist told her that the findings from the prenatal test showed that there was 'something wrong' with the chromosomes. When Mrs Roothaan asked her[1] about the exact nature of the problem, the physician said that things were a little too complicated to be explained over the phone. The chromosome analysis did not reveal an extra chromosome 21, as is the case with Down's syndrome, but some of the cells did show another extra chromosome. The exact significance was still unclear for the time being, but the consequences would in any case be far less serious than with Down's syndrome, the doctor immediately reassured her. A clear result would require further diagnostic testing, so the physician suggested an appointment at the hospital. In order to discuss the findings in full and consider the available options, Mrs Roothaan was to see her the next morning at nine, preferably together with her husband.

What news will be awaiting Mrs, and Mr, Roothaan in the consulting room of the clinical geneticist? When they decided to embark on this diagnostic trajectory they expected a clear, unambiguous result, either good or bad. But the way things have turned out, the findings seem to lie somewhere in between. The result is not really bad, as there is no evidence of Down's syndrome. Its possibility, after all, because of Mrs Roothaan's age, provided the principal reason for chorionic villi sampling. On the other hand, the result cannot be 100 per cent good, since they have to see the clinical geneticist.

In this case study[2], Mr and Mrs Roothaan are confronted with an ambiguous result: a finding that may or may not indicate a specific problem. Actually, it is a preliminary or 'grey' result, one of which it is still unclear whether it will turn out to be normal ('white') or aberrant ('black'). This kind of result is a common phenomenon in prenatal diagnosis. Chromosome investigations may produce a black result in line with the reason for doing the test (usually Down's syndrome), other black results, and also different kinds of grey results (see Box 1). In the daily practice of prenatal diagnosis, professionals do not apply a standard approach for dealing with grey results. The few publications on this subject seem to suggest that clarifying grey results is a mere technical matter.[3] This literature describes a number of technical procedures for the various types of grey results, but no mention is made of the fact that such approaches may also differ from one patient to the next (Van Zwieten, Willems et al. 2005). Entirely in agreement with the model of non-directive counselling that is applied within prenatal diagnosis, as in other clinical genetics settings, the recommended strategy for dealing with grey results is: 'To gather the best information available and to present it to the woman in a nondirective way. The aim is to allow her to make a fully informed decision which the counsellor then supports' (Kelly 1999, p. 183). This recommendation implies the assumption that 'gathering the best information available' is a strictly technical matter, and that the women involved (and their partners) only enter the equation when the process of 'gathering the best information available' has been completed, that is, from the moment that the grey result is communicated. In other words: it would make no difference for the clarification process of a grey result as such whether these findings concern Mrs and Mr Roothaan or Mrs and Mr Jones or Peterson.

But is this really the case? Is 'gathering the best information available' a purely technical matter? And does it make no difference to whom unclear prenatal diagnosis results have to be clarified? It is understandable that the literature on the subject presents a picture in which the result is not linked in any way to the individual involved. Due to the value-related implications of prenatal diagnosis practice, which may involve terminating a pregnancy, professionals constantly and explicitly stress their non-involvement in decisions once the findings have been presented. They refrain from any influencing of decisions that apply to the continuation of a pregnancy. The general picture of a result as purely

Box 1   Types of results of prenatal chromosome diagnosis

The majority (more than 70 per cent) of prenatal chromosome diagnosis takes place due to the related elevated risk of trisomy 21 (Down's syndrome). Apart from trisomy 21, other chromosome aberrations are regularly recorded.

Some of these chromosome aberrations are generally considered as **more serious than trisomy 21** and result in very severe physical and mental disabilities. The most commonly known aberrations are trisomy 13 (in which three instead of two chromosome 13 are present), trisomy 18 (three instead of two chromosome 18) and triploidy (three chromosomes instead of two – for all chromosomes).

There is another group of chromosome aberrations that is generally considered **less serious than trisomy 21**. Most common are the sex chromosomal aberrations, in which case there is something wrong with the x- or y-chromosomes. Among these, Turner's syndrome and Klinefelter's syndrome are quite familiar. With Turner's syndrome (45,x) we find one single x-chromosome instead of two sex chromosomes; with Klinefelter syndrome (47,xxy) there is one x-chromosome too many. Most phenotypical problems in Turner's syndrome and Klinefelter syndrome relate to infertility and/or secondary gender characteristics. In the case of a 45,x or 47,xxy appearing in mosaicism (see Box 2), these phenotypical problems are less serious.

In addition to the various types of aberrant results ('black' results), we can make a general distinction between two types of **grey results**, that is, mosaicism and structural aberrations

In **mosaicism** an individual shows two (or more) genetically different cell types. Forms of mosaicism have been observed, for example, of normal cells and cells with trisomy 21, or of normal cells and cells with an X- chromosome short, but mosaicism may appear in all kinds of variations.

Mosaicism is a relatively common phenomenon in chorionic villi sampling. As placental material is examined in chorionic villi sampling and not any foetal material, the observed mosaicism may occur in the placenta, but not manifest itself in the foetus. In that case one speaks of confined placental mosaicism. Chorionic villi sampling employs two different kinds of method to look at two types of placental cell (see Box 2 under the heading 'cell culture' for more information on short- and long-term cultures). When mosaicism is observed in the short-term culture of the chorionic villi sampling, additional testing can take place through a long-term culture. When mosaicism is *not* observed, there is an increased chance that mosaicism is limited to the placenta, than when it *is* found in the long-term culture.

In addition to mosaicism confined to the placenta, there may also be other explanations for mosaicism in laboratory material that cannot be found in the foetus. For instance, because genetically deviating cells have developed in the cell culture (culture arte-fact) or because not only foetal cells but also cell material from the mother has been examined (maternal contamination). Culture artefacts and maternal contamination can both occur in chorionic villi sampling as well as in amniocentesis.

In a **structural aberration** the problem concerns the form (structure) of the chromosomes. There are two kinds of structural aberration. With an unbalanced structural aberration, the form of one or more of the chromosomes has been altered in such a way that the total amount of genetic material has also changed. In a balanced structural aberration, the total amount of (chromo-somal) material has remained the same. Whereas an unbalanced aberration always leads to phenotypical aberrations, this is not commonly the case with a balanced aberration. To determine whether the chromosome aberration would lead to phenotypical aberrations, the chromosomes of the parents are analysed, as this may indicate whether the change in the chromosomes occurs within the family or is a new alteration. When one of the parents is a carrier of the same chromosome anomaly, it is expected that there will not be phenotypical aberrations; for the parent with the same structural aberration is of normal health. But when it con-cerns a new aberration, it cannot be excluded that the structural aberration will have phenotypical consequences.

---

technical information, with no link whatsoever to the person involved, perfectly matches the ideology of non-directiveness.

This does not discount the fact, however, that prenatal diagnosis actu-ally is a form of diagnosis. Furthermore, gathering information is never a random process; it always serves a specific purpose and takes place within a certain framework. As is true of all diagnostic formats, the information gathered to obtain a result will have to be assessed. It will have to be decided whether or not the findings are relevant in the light of a particular diagnostic context's stated purpose. And the person to whom this result applies is always intrinsically part of that context.

So how does the idea that a testing result consists of mere technical facts and the fact that prenatal diagnosis is a form of diagnosis in which the individual patient matters relate to each other? And how is this linked up with the ideology of non-directiveness? I will try to give some more insight into this matter by examining how professionals deal with grey testing results. I will examine, in particular, whether clarifying a grey result indeed involves a strictly technical process or whether, during that process, the specific patient involved is kept in mind as well.

Constructing results

The notion that prenatal diagnosis results involve straightforward technical issues has been challenged before. Rayna Rapp, a cultural anthropologist who has carried out an extensive observation study of the practice of prenatal diagnosis in the United States, demonstrated that assessment of information is an intrinsic part of the process of producing results (Rapp 1999). In laboratories where chromosomes are examined – described by Rapp as 'factories of fact construction' – technical information does not present itself unambiguously. According to Rapp, the work does not so much consist of observing and collecting facts, as it entails interpretation of ambiguous data. The daily responsibilities of laboratory staff include consideration of which data are of importance for a testing result and which are not:

> Learning to muffle background noise in favour of the normalizing grid is a central skill of their work. At the same time, they must highlight each entity which differs from the norm, investigating its potential to move from background insignificance to foreground significance.
>
> (Rapp 1999, p. 205)

Rapp describes professionals active in the field of prenatal chromosomal diagnosis as 'highly accomplished diagnostic disambiguators'. In other words, a result obtained through prenatal diagnosis is constructed information rather than something that 'presents itself'. And interpretation of technical information plays an important role in the construction process of the result involved.

The next question is what drives professionals in their interpretation of technical information? More particularly, what weight is given to individual patients in this process? As mentioned, I try to answer this question by specifically considering grey results. However, within this framework I view grey results not so much as exceptions to 'standard' (white or black) prenatal diagnosis findings, but as results marked by a blown-up disambiguating process as described by Rapp. A grey result, then, is a result in which this process is much more prominent than in 'straightforward' results, because in these cases the professionals explicitly question whether the technical information generated in the laboratory is eventually to be interpreted as normal (white) or abnormal (black).

In order to examine how professionals operate in clarifying grey results, I observed their daily activities at the Department of Clinical Genetics of the Academic Medical Centre (AMC) in Amsterdam. After becoming thoroughly acquainted with the health professionals involved and the way they work, my first step was to examine how a prenatal diagnosis result is normally arrived at in the laboratory setting. Special points of attention in this respect were the various contacts and consultation moments between the different professionals with their respective tasks, that is, the gynaecologist, lab technician, cytogeneticist and clinical geneticist (see Box 2).

Box 2   Constructing a result of prenatal chromosome diagnosis

Chorionic villi sampling is based on the fact that the cells derived from the placenta are genetically almost identical to the foetal cells. Examination of the chromosomes of the placental cells makes it possible to indirectly determine the chromosome pattern of the foetus. **Amniocentesis** makes use of the fact that the foetus, as of a certain moment in the pregnancy, secretes cells into the amniotic fluid. By examining these cells, the chromosome pattern of the foetus may be established.

In order to be able to carry out chromosome testing, **the cells to be examined must first be extracted from the body of the pregnant woman**. This is done by a **gynaecologist** who, via the vagina or with a hollow needle through the abdomen, retrieves some placenta tissue or amniotic fluid from the woman's body. There is a chance of around one per cent that a miscarriage is initiated through this invasive procedure (Health Council of the Netherlands 2004).

Consequently, a **laboratory technician** processes the testing material in a number of steps. With chorionic villi sampling the

---

foetal material is first **dissected** from any other body material that is extracted during the procedure.

The next step is **increasing the number of available cells**. In the case of chorionic villi sampling the manner of increasing the number of cells **(cell culture)** depends on the analysis method applied. In the fast method – the short-term culture – the results are quickly available, but less reliable as the examined cells are not always genetically the same as the foetal cells. The results of the long-term culture are more reliable, but this method is more time-consuming. The most reliable results are reached through a combination of the fast and the slow methods, but this combination not only takes considerable time, it also requires a lot of work. This is the reason why the AMC makes use of an alternative developed at the hospital (Schuring-Blom 2001). Part of the placental material is directly used for the short-term culture and the remainder – if any – is stored. Only when an aberration is found in the short-term culture will the remaining material be used for a long-term culture. In the case of amniocentesis, a cell culture takes just about as long as the long-term culture of chorionic villi sampling (seven to ten days).

The technician performs the following actions **to make the chromosomes visible**. Once the cells have been manipulated to the division phase in which the chromosomes can be discerned optimally, the testing material is applied to a slide, which, after colouring, can be viewed under the microscope.

Finally, the technician notes the microscope image in a **schematic representation of the chromosomes**. This is done in two different ways. First, the number of chromosomes of ten observed cells is recorded on a specially designed counting form. In order to optimize the chances that the schematic representation matches the foetal chromosome pattern, it is standard procedure for technicians to minimally count ten cells stemming from two different slides. In the case of one aberrant cell being found in the ten cells that are counted, the number of cells for examination is increased. In the second place, a schematic representation is made of all chromosomes of each of two cells. In such a diagram – a karyogram – the 22 chromosome pairs and the x- and y-chromosomes are arranged in a fixed order. To this end a photograph is taken of the microscope image. All chromosomes are – literally – cut from this picture and pasted on a form by the technician in the prescribed order. Nowadays, a computer program is used to produce a karyogram.

> The **chromosome diagrams** made by the technician are **summarized in a karyotype** by the cytogeneticist. The cytogeneticist checks the written material supplied by the technician and summarizes it in the karyotype, a standard notation of the chromosome pattern of one individual, in which 46,XY indicates a normal male and 46,XX a normal female karyotype.
>
> After consultation between the cytogeneticist and clinical geneticist – who in the AMC work in the same department, contrary to customary practice in most other centres – the found **karyotype is phrased as a result**, for instance 'normal male karyotype'.
>
> The **result letter** states the three last-mentioned steps. Under the heading 'examination findings' the exact number of analysed cells is listed with accompanying chromosomes. Under 'karyotype' the karyotype is noted in the prescribed manner, for example, 47,XY,+21. Under the heading 'Conclusion' the karyotype is phrased in plain words, for example, 'male foetus with Down's syndrome'. Under 'Comments' there is room for various matters that are (or may be) of importance for a correct interpretation and/or communication of the result; for instance, whether, and in what way, the parents have already been informed of the result. Any technical details of the analysis can also be included. In the AMC both the cytogeneticist and the clinical geneticist sign the result letter.

The actual observations concerned the interdisciplinary consultations between the professionals during the 'construction' of a result. Based on Rapp's notion of 'disambiguating process', this should be visible during consultations between professionals and therefore open for study. Accordingly, I agreed on a procedure with the professionals involved to ensure that I could be present in the department for observation at key moments.[4]

The series of professional actions during which a result is constructed is conceptualized as a 'testing trajectory'. From the perspective of the parents who take part in prenatal diagnosis, this testing trajectory starts earlier and continues after that point. Within the framework of this study geared toward the clarification process of grey results, I zoom in on the trajectory from the intervention by the gynaecologist to the drawing up of the final result. The observations were to answer the following question: What do professionals do to clarify grey results, that is, to make them black or white?

My observations, done from July to September 2001 and from April to May 2002, were reported in a so-called observation protocol.[5] To allow for analysis of the observations,[6] all documented and monitored result trajectories were considered as case studies, which were studied individually as well as in relation to each other. This analysis aimed at providing an answer to the above-mentioned question, but I also analysed the material in the light of the following, broader question: Does the way in which professionals deal with grey results correspond with the general idea of a testing result as a purely technical matter?

Clarification of grey results in practice

A little more than half (22/39) of the testing trajectories I observed produced an aberrant result in the manner described in Box 2. In these 'straightforward' trajectories, the inter-professional consultations were not related to the result itself, but applied to more practical matters as 'this result should be completed a.s.a.p.' or 'who will inform the couple of our findings?'

In slightly fewer than half of the observed testing trajectories (17/39), a grey result was generated at some point along the trajectory. The approach selected by the professionals to clarify these grey results – that is, to make them white or black – corresponded with the literature on the subject (Van Zwieten, Willems et al. 2005). The following summary describes the different kinds of follow-up procedures applied in the observed testing trajectories that were aimed at clarifying the initial grey results.

The trajectories in which the grey result concerned mosaicism (14/17) mainly focused on the question whether the noted chromosome aberration would also occur in the foetus, or, in other words, whether the findings of the chromosome analysis were (sufficiently) representative for the foetal chromosome pattern (see Box 1). Additional testing and analysis in the course of this aimed at verifying whether the chromosome aberration could be found in other types of cells as well. If so, the likelihood of the aberration also manifesting itself in the foetus would become greater. In principle, mosaicism found in the short-term culture of chorionic villi sampling was followed by a long-term culture (see Box 2, under cell culture). In most cases where there was insufficient material available, a new invasive test in the form of amniocentesis was suggested to the parents. Additional investigation of the chromosomes in other types of cells thus allowed for turning a grey result into either a white one or a black one.

Both possibilities were found in the observed testing trajectories: some grey preliminary findings became black results and others changed into white results. For instance, a grey result (mosaicism trisomy 21, i.e., a combination of a normal cell line and a cell line with trisomy 21) recorded in the short-term culture of the chorionic villi sampling turned black once a full trisomy 21 was found in the long-term culture. Another grey result (mosaicism of normal cells and cells with each having three extra chromosomes: 2, 6 and 7) became white once mosaicism was not confirmed in amniocentesis. The fact that an aberration was found in the (short-term culture of the) chorionic villi sampling that would not occur within the foetus itself was explained in this specific testing trajectory by regarding the original chromosome aberration as mosaicism confined to the placenta (see Box 1).

In additional analysis, alternatives to chromosome testing were sometimes applied. In one trajectory this concerned an ultrasound scan; in this particular case the phenotype of the foetus was examined instead of the genotype. The grey result consisted of mosaicism in the amniotic fluid – a much rarer phenomenon than mosaicism in chorionic villi sampling. Because it involved a type that is rarely found in humans (46,XY/46,XX: a combination of normal male and healthy female cells), the cytogeneticist as well as the clinical geneticist considered it unlikely that it would materialize in the foetus. The assumption therefore was that the chromosomes found in the laboratory were not representative for the foetal chromosomes. In other words: the grey result should actually be white. However, this required an explanation as to why mosaicism had been observed in the laboratory. The first possible explanation was that the preparations of two patients of different gender had been mixed up. This implied that only XX-cells could have been found in one culture and that the other culture exclusively contained XY-cells. This was not the case, however, as the combination of XX- and XY-cells had been observed in two separate cultures. A second explanation was the possibility of maternal contamination: all XX-cells would then be derived from the mother and all XY-cells would originate from the foetus. Additional testing in this specific trajectory was needed to support this hypothesis. An ultrasound scan was performed to determine whether the foetus had male genital organs. If so – reasoning back to the genotype – it could be assumed that the tested XY-cells definitely derived from the foetus. The ultrasound scan showed that the foetus indeed was male. And as the scan revealed no further phenotypical aberrations in the foetus, it was concluded that the XX-cells did not originate in the foetus, but were derived from its mother. By testing the actual phenotype of the foetus, the explanation of maternal contamination was thus confirmed and the preliminary grey result turned white.

In the remaining (3/17) testing trajectories, the grey result consisted of a balanced structural aberration. In these trajectories, the professionals did not opt for testing other foetal cells in order to clarify the results. Additional testing and analysis focused on the chromosomes of the parents instead. With a balanced structural aberration, it is unclear whether the chromosome aberration (genotype) observed will have phenotypical consequences (see Box 1). The result in all three trajectories was clarified by testing the chromosomes of the parents. In each case the chromosome aberration of the foetus was found in one of the parents as well, the conclusion being that they were dealing with a familial structural aberration in which no phenotypical consequences were to be expected. Additional chromosomal examinations of the parents thus made these grey results white.[7]

Hesitation in clarifying grey results

The observations of grey testing trajectories revealed a remarkable phenomenon: in four trajectories (always with mosaicism as a preliminary outcome) clarifying the result was not considered the obvious next step in the process. In these cases, there was much hesitation among those involved on how to proceed, which gave rise to more professional interactions, mutual consultations and lengthy discussion of cases at weekly meetings, and so on. This hesitation with respect to clarifying initial findings was not based on technical matters; for instance, because it was unclear which approach to adopt. It seemed rather to stem from the question whether the current ambiguity actually *needed* clarification. While reflecting on whether clarification of a grey result was (still) necessary or desirable, the professionals also took future decisions on pregnancy continuation or termination into account. Because the idea of clarifying a grey result as a purely technical affair did not seem to hold in these four cases, I will describe them in detail and indicate the specific reasons that prompted the professionals involved to raise this issue. In these descriptions I will show how in a grey result's clarification process technical issues and aspects related to the individual patient were both influential.

Clarifying a grey result in the case of a fixed decision to continue the pregnancy

In the case of a woman who had had chorionic villi sampling due to an ICSI[8] pregnancy, and enlarged nuchal translucency,[9] the diagnosis was mosaicism of normal cells and cells with 45,x (mosaic for Turner's syndrome: see Box 1). Of the 17 analysed cells, two showed a single x-chromosome and all the other ones had a normal 46,xx chromosome pattern. These findings constituted a grey result as the question remained whether the foetus would also show Turner's syndrome mosaicism (or that it would concern mosaicism confined to the placenta, see Box 1). The cytogeneticist, whose job it was to summarize the chromosome diagrams drawn up by the technician in a karyotype (see Box 2), decided for a long-term culture to clarify the result.

So far, the testing trajectory had gone along the same lines as when, for instance, a mosaic trisomy 21 found in the short-term culture was followed by a long-term culture. In this case, however, the professionals were dealing with mosaicism of a sex chromosome abnormality, with symptoms that were much less severe than those of Down's syndrome. This was partly the reason that this testing trajectory changed after informing the parents of the outcome of the short-term culture. It was explained to them that the long-term culture's final result would take quite some time. The parents, however, immediately indicated that they would not terminate the pregnancy now that Down's syndrome had been ruled out. For the professionals, this new situation raised the question whether the result still needed clarification. The observation demonstrates how the issue was brought up in the weekly interdisciplinary meeting, when the clinical geneticist reported on her contact with the parents:

'With respect to this mosaicism, a thick nuchal translucency was observed, while both 46,xx-cells and two 45,x-cells were found in the chorion. We may be dealing with Turner. I phoned the parents but they were merely very pleased that their child did not have Down's syndrome; a girl with Turner would be quite acceptable to them.' One of the gynaecologists gives a somewhat worried response and wonders if the parents have a clear picture of Turner's syndrome. She mentions all uncertain factors with respect to the current result. The cytogeneticist replies: 'That may be the case, but even when I find more cells with 45,x, it won't change my opinion as the result is acceptable to them.' The gynaecologist remarks that she only hopes that the parents don't take the situation too lightly, whereupon the clinical geneticist reacts: 'It is an ICSI pregnancy, and therefore a very costly one ...'

After the weekly interdisciplinary meeting, I asked the cytogeneticist for further comments. She explained the situation as follows:

We initiated a long-term culture with three possible outcomes: 46,xx [normal girl], 45,x [girl with Turner's syndrome] or 46,xx/45,x

[Turner's syndrome mosaicism]. But even if the findings are not good, no additional testing will be carried out, as the parents have indicated they will not terminate the pregnancy.

Although the professionals considered the result of the short-term culture a grey result, needing clarification, it was decided not to proceed with the clarification process. After all, the parents had made it quite clear that they would continue the pregnancy anyway; to them, the initial outcome needed no further clarification.

In the case of a woman who had had chorionic villi sampling on the basis of a DNA diagnosis,[10] the technician observed 45,X with three of the 16 cells counted, while the others were normal 46,XY cells. This was a grey result because it was not clear if the 45,X cells would be present in the foetus or not. A long-term culture could not be performed because all material had been sent out for DNA analysis. The observation protocol states the following on this result:

> The cytogeneticist and clinical geneticist discuss how to proceed. CG. [clinical geneticist] says that she will call the parents tonight to tell them that she has good news on the DNA analysis but that the outcome of the chromosome test is still unclear. She will also explain that there are two options, i.e., an amniocentesis or an ultrasound scan to determine whether it is possible to see normal male genitals.[11]

At the weekly interdisciplinary meeting the clinical geneticist reports as follows:

> Early this week I heard that the child does not suffer from the hereditary metabolic disorder. We have examined the chromosomes and found mosaicism. In the past we followed up with amniocentesis. I have discussed this with the parents on the phone but they do not agree to an amniocentesis. I explained to them that there is a 90 per cent chance that there is no problem at all, or a 10 per cent chance that something is wrong with genital formation. Well, this was entirely acceptable to them. When she told her husband, he replied that they were more than willing to take their chances in this matter. They told me that they knew a child that suffered from a similar problem with its genital organs. It turned out to be operable and, moreover, in relation to their child's possible problems [hereditary metabolic disorder, the initial reason for prenatal diagnosis] these issues were no obstacle at all for continuing the pregnancy.

The position of these parents led to a similar situation as in the first case. However, in the absence of material, in this second case a long-term culture was either impossible or would require a second invasive test, with a new risk of miscarriage. In both cases the parents played a role in establishing the relative importance of further examination of the chromosome aberration. They also assessed the seriousness of this possibility and related their assessment to their personal situation. But whereas the first trajectory's personal context was primarily determined by the 'cost' of the ICSI pregnancy, problems in the second case were related to the couple's original worries, that is, their difficult experiences with the metabolic disorder from which their first child was suffering. In light of these circumstances, the parents did not favour additional analysis, a feeling that grew even stronger when learning that a second invasive procedure came with a renewed risk of miscarriage. They were not prepared to take this risk and declined further clarification of the grey result, at least for the time being.

Clarifying a grey result in case of an assumed decision to terminate the pregnancy

In the testing trajectory of a woman who had had amniocentesis in view of her age, trisomy 8 mosaicism was observed, whereas the technician had already found a trisomy 21. The trisomy 8 was recorded in two cells, which raised the question whether trisomy 8 mosaicism would be representative of the foetus. When a lab technician observes that two of the ten counted cells are aberrant, it is normal procedure that she goes on to analyse more than ten cells (see Box 2, schematic chromosome representation). But the technician who found this trisomy 8 mosaicism, after already observing a trisomy 21, apparently took the possibility into account that this specific grey result needed no further clarification and put the matter whether or not to count more than ten cells to the cytogeneticist. The entry in the observation protocol of the meeting between the technician and the cytogeneticist states the following:

> C. [cytogeneticist] says to T. [technician] that she will consult CG. [clinical geneticist] whether it is of importance to clarify the trisomy 8 as well. T. remarks: 'Normally speaking, we continue the analysis with a trisomy 8.' 'True', C. replies, 'but things may be different in this case because of the trisomy 21. So I am not sure that you should go the extra mile. And the trisomy 8 may be nothing more than a culture artefact; actually that's what I suspect. So finish those ten cells first, and then we will decide what to do next.'

The observation protocol mentions the following on the consultations between the cytogeneticist and the clinical geneticist:

> CG. asks: 'What do you normally do, that is to say when trisomy 21 is not found?' C.: 'We would continue looking, but I am not sure if this is the right option in this particular case, in view of our busy schedule as well.' CG.: 'I suspect that the pregnancy will be terminated anyway and this would make it a waste of time indeed.'

Just as in the previous cases the question arises whether the grey result (trisomy 8 mosaicism) should actually be clarified. This time, however, the question is not raised after, or in response to, the contact with the parents, but originates with the professionals themselves. Moreover, in this trajectory the reason for hesitation is exactly opposite to that in the situation in the previous trajectory. This time, the professionals believed that the pregnancy would be terminated anyway, based on the observed trisomy 21. Therefore they considered information on a possible trisomy 8 mosaicism as no longer relevant. Consequently, there is no mention of the mosaicism trisomy 8 in the results letter.

> Findings of the analysis: structural chromosomal aberrations: no recognizable (Q-banding)
> Karyotype: 47,XX,+21
> Conclusion: female foetus with trisomy 21

Clarifying a grey result in case of a fixed decision to terminate the pregnancy

This case concerns a woman who had an amniocentesis due to her age and a raised risk based on the triple test.[12] Seven cells 47,XXX and three cells 45,X were observed: not mosaicism of a normal and an aberrant cell line (as is usual), but mosaicism of two aberrant cell lines. The cytogeneticist decided to clarify the grey result by applying an advanced analysis method, i-FISH,[13] on the already available amniotic fluid. This method enabled her to gather more information on the number of aberrant cells, the advantage being that a second invasive test – with the risk of miscarriage – was not required. In consultation with the clinical geneticist it was decided to include an ultrasound scan in the follow-up trajectory. This would provide data on the sex of the foetus and any heart abnormalities, which might be an indication of the presence of Turner's syndrome (45,X).

After the start of the additional testing, the clinical geneticist met with the parents to inform them about the preliminary findings and the options open to them. In the weekly interdisciplinary meeting after the clinical geneticist's appointment with the parents, she reported that it looked as if the parents would probably continue the pregnancy. But in the following week this trajectory took a surprising turn. The observation protocol:

> I call C. to find out if she has any information yet on the i-FISH. C. replies: 'No, but CG. has pointed out in the meantime that the parents have decided over the weekend to terminate the pregnancy, rather unexpectedly I might say.'

The observation protocol on the weekly interdisciplinary meeting:

> CG. explains the case of the woman with mosaicism karyotype for whom a FISH has been performed: 'Last week I talked to the parents and explained that it is difficult to come up with a reliable prognosis. I told them about the two possible phenotypes and mentioned the possibility that the ovaries might hardly develop, implying possible infertility. I also mentioned the website of our Danish colleagues. It has a lot of balanced information that might be of importance to them. At the same time I warned them not to start looking for info on the internet on their own. The possible symptoms are presented as far too serious and I pointed out to them why these would not be relevant for their future daughter. On Monday, the parents saw G. [gynaecologist] whom they told they had decided to terminate the pregnancy, in part because all the information on the net about Turner had really scared them. Well, you can imagine I was not very happy with what had happened. But then again, I simply had to tell them about the website … '

While the professionals were still working on clarifying the grey result, the parents had already decided to terminate the pregnancy. This implied that there was no longer any reason for the professionals to follow-up on the initial grey findings. But in this particular instance, things even went a little further. Further interactions revealed that the clinical geneticist did not even want to clarify the grey result.

> After a remark by someone else, CG. says: 'I have heard nothing else about it, but yes, G. made it quite clear that they would terminate. In that case we will have a skin biopsy. And in my opinion a post-mortem

examination of the ovaries is undesirable, because if that turns out normal ... '

D. [doctor-sonographer]: 'you just don't want to know ... '

CG: 'No.'

D: 'The same goes for the heart, if that is normal ... '

CG: 'No, and there is also no relevance for the follow-up; other forms of mosaicism have their own characteristics. I must say that I am very disappointed about it all, but what can I do ...?'

Because the parents have already decided to terminate the pregnancy, the grey result does not need any clarification, there is no benefit in pursuing certainty. In this case the professionals even pro-actively reject any clarification because when the heart and the ovaries turn out normal, this would indicate a healthy foetus without Turner's syndrome. Of course, it would be quite awkward to learn that the foetus did not have the suspected chromosome abnormality after all. The statement 'you just don't want to know' positively indicates that any additional information has become 'undesirable', not only for the parents but also for the professionals.

Technological and moral work interlinked

Regarding the question what professionals do to clarify grey results, my observation suggests that in terms of technical procedures the clarification process is in line with the literature, whereby a distinction is made between the two most common grey results, mosaicism and balanced structural aberrations (Van Zwieten, Willems et al. 2005). But, remarkably perhaps, some preliminary findings prompted the question whether a grey result needed clarification in the first place. This question brought into focus the individual patient involved. In all four cases there was doubt on the relevance of the outcome of the chromosome tests. The main issue was whether the information of the result would (still) be of importance for the parents' decision about the pregnancy – a question that was mostly answered by the parents, but in one case exclusively by the professionals.

The descriptions of these testing trajectories go against the idea that clarifying a grey result involves a strictly technical process in which the individual patient plays no particular part. Naturally, the moment when the professionals consider future decisions on terminating the pregnancy, the individual patient appears – literally or figuratively – on stage. While still in the phase of constructing a testing result, then, in some of the observed cases the professionals did not focus on technical matters only, but also took the individual patient into account, particularly their – assumed – considerations regarding termination of pregnancy.

While my main concern was to find out how professionals deal with grey results, I also wanted to examine whether the observed practice of clarifying a grey result would correspond with the general idea of a testing result as a purely technical matter. As indicated, I defined a grey result within the framework of this study as a result where it remains unclear whether it will ultimately turn out normal ('white') or aberrant ('black'). In case of mosaicism, the question was whether the observed chromosome aberration was representative of the foetal chromosome pattern. The grey mosaicism became a black result when it was assumed – based on further testing where possible – that the chromosome aberration found in the lab was representative of the chromosome pattern of the foetus. In the case of balanced structural aberrations, the question was whether the observed chromosome aberration would have phenotypical consequences. The grey structural aberration became a black result when it was assumed – based on further testing where possible – that the noted chromosome aberration would have phenotypical consequences (with a varying degree of seriousness).

What, then, should be inferred from the fact that in some cases of grey results, professionals doubted whether the chromosome aberration was actually relevant in the light of the decision on the continuation of the pregnancy? The logical conclusion would be that professionals apparently only considered clarifying a grey result in prenatal diagnosis when it was assumed – either by the parents or by the professionals – that the black outcome would be relevant for deciding on terminating the pregnancy. It seems, then, that generally professionals do not only view an aberrant result as information on chromosomes that is representative of the foetal genotype and as information on chromosome aberrations with clinical consequences; they also tend to view an aberrant result in prenatal diagnosis as information which is relevant for the decision on continuing or terminating the pregnancy involved.

How does this conclusion relate to the idea that a testing result is a purely technical matter? Whereas assessments regarding the representativeness and phenotypical consequences of chromosomal information can be considered as strictly (medical-)technical matters, assessment regarding the relevance of chromosomal information for a decision about the termination of a pregnancy cannot be considered as such. Instead, this assessment clearly has a moral dimension. Significantly, then, my observations challenge the general idea of a testing result as a strictly technical matter. In the four cases my observations were at odds with this standard view because, in the process of disambiguating results, the question regarding the continuation of the pregnancy came up before the final results were in, and it even determined the extent to which results were actually disambiguated. This means that the technical work of producing a result (usually attributed to the professional) and the moral work of making a decision about the continuation of the pregnancy (usually attributed to the patient) were more interlinked than is generally assumed. Moreover, while the model of nondirective counselling is applied to prevent any involvement in moral decisions based on the testing result, this chapter shows that moral involvement can also occur in the process of making testing results.

New professional responsibilities

The conclusion that the moral aspect of decisions on either continuing or terminating a pregnancy is an intrinsic part of prenatal diagnosis is significant for at least two reasons. First of all, it is important for communication with parents. Although professionals are quite reluctant to openly discuss the option of abortion with pregnant women in the early stages of prenatal diagnosis (Press 2000), the empirical results underscore that it is not necessary only to refer explicitly to the issue of deciding on the termination of pregnancy when a black result shows up. After all, my observations show that in the case of an indefinite grey result, the parents can also be prompted to reflect on what this result means to them, which they can only do in relation to their decision on termination of pregnancy. More details of this kind of communication with parents about grey results are discussed elsewhere (Van Zwieten, Willems et al. 2006), but here I stress the professionals' responsibility to inform parents about the possibility of their personal ideas about continuation or termination of pregnancy influencing the course of an indefinite testing trajectory. More generally, it is never too early to emphasize why prenatal testing results are generated in the first place. This applies to prenatal screening as well. Although promoting informed choice is commonly recognized as the chief purpose and benefit of prenatal screening (Williams, Alderson and Farsides 2002), the freedom-of-choice argument is not always specified. Therefore, prenatal counselling should, in my view, always explicitly discuss the fact that prenatal testing results are meant to serve as input for the parents' decision about continuation or termination of pregnancy.

Secondly, the conclusion that professionals, while constructing results, sometimes take decisions on pregnancy termination into account also has a bearing on policy issues. Our empirical results challenge an essential foundation of the ideology of nondirectiveness, namely the implication that prenatal diagnosis is a value-free form of information gathering. In this respect, Biesecker claimed that 'An insistence on nondirectiveness has stymied the process of policy making in prenatal genetic testing ... In the name of nondirectiveness, genetic counsellors have avoided their professional and moral obligations to take a stand on the appropriateness of certain types of prenatal testing' (Biesecker 2000, p. 980). Could this analysis be true? Let us hope not. Due to technical developments, it is not at all clear what the world of prenatal diagnosis will look like in the near future. There is an increasing range of genetic abnormalities to screen for, as well as increasing possibilities of excluding some of the grey results in a scenario of 'targeted testing' (Ogilvie 2003). In this turbulent situation, I believe the providers of these services should wholeheartedly participate in the discussion about which tests should and should not be offered. Their refraining to do so would indeed be a neglect of their professional responsibility. For such an attitude would not correspond with their actual involvement in the daily practice of prenatal diagnosis. This study demonstrates that a prenatal diagnostic result is not merely information for professionals, but information they consider relevant for the parents' decision either to continue or terminate the pregnancy. Because occasionally the construction of testing results involves such a moral dimension, professionals should not refrain from, but participate in, public-ethical discussions about prenatal testing.

Acknowledgements

This research was part funded by the Netherlands Organization for Health Research and Development (ZON MW). Supervisors in this study were Dick Willems and Nico Leschot, and Lia Knegt for the technical aspects. I would like to thank all staff of the AMC Prenatal Diagnosis Department who have so generously helped me within the framework of this practice survey. I also wish to thank Wilma Poelma for her extensive explanations of all laboratory procedures and Heleen Schuring-Blom for checking the technical terms.

3
Genetic Diagnostics for Hereditary Breast Cancer: Displacement of Uncertainty and Responsibility

Marianne Boenink

Questioning the language of autonomous decision-making

> Women from families with frequent breast cancer can have their DNA tested. When the gene involved proves to have a defect they can decide to have their breasts removed preventively.
>
> *(de Volkskrant,* 15 May 2004).

It sounds quite straightforward: women who suspect a high occurrence of breast cancer in their family simply do a genetic test that predicts their chance of contracting it. As such, predictive DNA diagnostics seems comparable to doing a blood test in order to establish anaemia or a urine test to identify a bladder infection.

Unfortunately, things are a little more complicated in the case of breast cancer. For one thing, the test involved does not determine whether you are ill or not, but whether you will fall ill at some point in the future. Predictive diagnostics thus raises the issue whether women have any desire to know that they have an increased risk of breast cancer. If so, what can they do with that information? In the case of a genetic defect for breast cancer anyway, the options for prevention are limited and drastic: either regular monitoring or preventive mastectomy. In particular the question of removing healthy tissue to prevent a future illness is a serious issue that has received much attention in the media.[1]

In the practice of DNA diagnostics these tough questions are often answered with reference to the autonomy of the client. Professional geneticists usually feel that it is the client who must decide whether she wants to have DNA diagnostics and how she wants to respond to the potential results. Their own role consists of providing information, explaining the various options and elucidating their implications; they refrain from normative judgments on the decisions taken by their client.[2] For instance, the brochure entitled 'Hereditary breast cancer and ovarian cancer', issued by the Dutch Cancer Society, emphasizes the client's autonomy: 'you decide for *yourself* whether you want to be tested to find out if heredity plays a role' (Dutch Cancer Society, 2002; my emphasis).

The thinking in terms of autonomy is tied to two assumptions that deserve further scrutiny. First, there is the assumption of a fully developed test or technique that is ready to be deployed and that it is unproblematic. Starting from this assumption, ethical reflection on predictive DNA diagnostics applies to the choices someone has to make before and after taking the tests, while the technology itself does not become an object of reflection.[3] Second, emphasis on the client's autonomy implies a specific distribution of roles and responsibilities between the clinical geneticist and the client: the geneticist provides relevant information in a neutral way, after which the client weighs that information and decides what to do. In other words, the thinking in terms of autonomy presupposes a technology that is ready for use and a client who takes the responsibility for choices with respect to such use.

Over recent years much has been said and written on the tenability and desirability of the autonomy ideal in the context of DNA diagnostics. For instance, the notion that geneticists can provide information in a neutral and nondirective way has been criticized on empirical as well as conceptual grounds (Clarke 1991; Grandstrand Gervais 1993; Mitchie, Bron et a1.1997; Rentmeester 2001; Steendam 1996; Williams, Alderson et al. 2002; Williams, Alderson and Farsides, 2002; Van Zuuren 1997). Another point of critique is that, by definition, genetics involves a family affair, meaning that a decision of one individual to have herself tested automatically has implications for other family members (Bolt 1997; Huibers and Van 't Spijker 1998; Husted 1997; Niermeijer 2004). My commentary is geared to the two assumptions of the thinking on autonomy mentioned above: the nature of the test and the distribution of responsibilities related to it.

In this chapter, based on observations of clinical-genetic practice and interviews with those involved, I will reconstruct the practice of predictive DNA diagnostics for hereditary breast/ovarian cancer as a lengthy and complex trajectory, in which uncertainty is not reduced, but constantly displaced and transformed.[4] This I will illustrate on the basis of three elements: pedigree analysis, DNA testing and the ways in which decisions are taken. My argument will reveal that there is no ready-made predictive DNA diagnostic 'test' for breast/ovarian cancer, but only a test-in-the-making that is incapable of generating results on its own. For this diagnostics to supply relevant results, clients have to fulfil an active role before and during the test, yet this particular task conflicts with the autonomy model.[5] Thus they are not simply users of a ready-made test; in order to function, the test-in-the making is dependent on the active input of clients. By pursuing this line of reasoning, I will show that clients are made co-responsible for muting, and thus minimizing, the uncertainties that belong to the 'test'. By foregrounding the technology's practice, it becomes clear that such a distribution of responsibilities raises various normative issues. Is it acceptable that clients have to compensate for the technique's shortcomings? Is the language of autonomy in fact appropriate for articulating those responsibilities? Before addressing the practice of DNA diagnostics, I briefly discuss the current level of knowledge on hereditary breast/ovarian cancer.

Hereditary breast/ovarian cancer

In Western countries like the Netherlands, breast cancer has a high level of occurrence. Only a small portion of all cases of breast cancer, however, has a hereditary basis. In genetics a distinction is made between hereditary, familial and sporadic breast cancer. Hereditary breast cancer requires at least three first-degree family members (or second-degree in the case of paternal heredity) within at least two consecutive generations to have (had) breast cancer, of whom at least one patient was under the age of 50 at the time of diagnosis. It is estimated that 5 to 8 per cent of all breast cancer cases involves this hereditary form. Familial breast cancer pertains to breast cancers that are discovered at a relatively young age or that involve at least two first-degree family members (or second-degree, in case of paternal heredity), but that do not meet the definition of hereditary breast cancer; this applies to about 15 per cent of all breast cancers (STOET 2001). All other cases of breast cancer count as sporadic.[6]

Today two genes are known to contribute to the risk of breast cancer: BRCA1 and BRCA2. A myriad of mutations has meanwhile been identified in both genes.[7] Those who inherit such a mutation do not necessarily contract breast cancer. The penetration of the mutation is incomplete: a mutation carrier has a chance of 50 to 85 per cent that at some point in her life a woman will contract breast cancer, while for the average Dutch woman this risk is 10 to 12 per cent. But the chance that a mutation carrier will contract ovarian cancer is increased as well, which is why one commonly refers to the combined notion of 'hereditary breast/ovarian cancer'.

In the case of a BRCA1 mutation, the risk of ovarian cancer is 40 to 60 per cent, and in the case of a BRCA2 mutation 15 to 20 per cent. In men with a mutation at BRCA2 the risk of breast cancer is also somewhat higher (6 per cent), while a mutation at BRCA1 appears not to increase men's risk (Dutch Cancer Society 2002, p. 10).

Today in the Netherlands, DNA analysis implies the taking of several blood samples from a person, after which both BRCA genes are completely analysed in the lab for deviations. It takes some six months for results to become available. If a specific mutation has already been identified in a family, it suffices to do a targeted analysis of the mutation involved, which generally takes two months. The mutations at BRCA1 and BRCA2, however, only explain a limited percentage of all cases of hereditary breast/ovarian cancer: today, a BRCA1 or BRCA2 mutation is detected in an estimated 25 per cent of families with a pattern that suggests hereditary breast/ovarian cancer (Dutch Cancer Society 2002, p. 14). This implies that other genes may be involved as well, or that there is a complex inter-action among genes or between genes and their environment. As a result, the predictive value of today's DNA diagnostics is limited; in many cases no deviations are found, while the family history suggests grounds for assuming a hereditary risk.[8]

If there is an increased risk, there are several possible preventive meas-ures. First, it is possible to have regular monitoring of the breasts (from age 25) and ovaries (from age 35). The first happens in a semi-annual breast examination by a surgeon, a yearly mammography (an X-ray of the breasts) and sometimes an ultrasound, while the second involves a yearly internal examination by a gynaecologist, a vaginal ultrasound scan and a blood test. The other option is surgery, whereby breasts and/or ovaries are removed preventively. In both cases there will still be a (slight) chance of cancer in the remaining breast tissue and/or the abdominal cavity. In the case of breast removal, concurrent breast reconstruction may be performed. In the case of removed ovaries, a woman can no longer have children and she will enter menopause prematurely. For this reason, and also because the risk of ovarian cancer before the age of 40 is still low, it is often suggested leaving such surgery until after that age (Dutch Cancer Society 2002, pp. 15–17).

In the Netherlands, DNA diagnostics is exclusively performed in a dozen Clinical Genetic Centres, usually linked to academic hospitals. The nor-mal procedure starts with an intake by telephone. If the client seems a likely candidate for hereditary breast/ovarian cancer, she is invited for a consultation with a clinical geneticist. In the subsequent predictive (or pre-symptomatic) diagnostic process, according to the Dutch geneticists

Leschot and Brunner, the following phases of genetic counselling can be identified:

- Information on the disorder, risks and testing options; estimating the psychological effects, and discussing the intervention options in case of a positive test result.
- Taking blood samples, making arrangements on test results and follow-up.
- Test result; deciding about subsequent steps; further information and additional clinical testing.
- Follow-up consultations: does the client understand the message, is she able to cope with it, is there a clear plan for the future?

(Leschot and Brunner 1998, p. 5)

Clearly, this phase model reflects thinking in terms of autonomy and its assumptions. The technology is considered a given; the clinical geneticist provides information on the technique, the results and the various options; and the client ponders, weighs the pros and cons, and makes decisions.

Now that DNA diagnostics for hereditary breast/ovarian cancer is not as new anymore and more experience has been gained with counselling, these phases are not as strictly adhered to in the Netherlands.[9] Counselling by psychologists or social workers especially has become less extensive; in the case of BRCA-testing, basically it is provided only when the client asks for it. Follow-up consultations with the clinical geneticist in the wake of the consultation on test results have become rare. But even with this provision, the practice of predictive diagnostics for breast/ovarian cancer proves to be much more complex than the above model suggests. As I will argue below; in particular, assumptions about the technology and the distribution of tasks between geneticist and client need to be put into perspective.

'Doing homework': pedigree analysis

Predictive DNA diagnostics is not just a matter of taking blood samples, doing a lab test and communicating the result, this becomes clear when someone contacts the clinical genetics policlinic for an appointment. Contrary to some of the interviewed clients' expectations, a date and time for a consultation is not immediately arranged, not even in cases of referral by a family physician or a surgeon. The clinic itself makes a selection by asking for family data on the phone. Based on this information,

clinic staff consider whether there is sufficient reason to suspect hereditary or familial breast/ovarian cancer. If so, the client will receive a letter
with a proposed date for a consultation with a clinical geneticist.

The first counselling interview is only in part about the client herself
and DNA diagnostics, for it mainly addresses her family. Those who get
an appointment also receive a form on which they have to enter the
dates of birth and death of the members of their family's last three generations, as well as the histories of their illnesses. A client who does not
have all relevant information is advised to inquire of her relatives. She
should bring along the filled-out form to the first appointment. 'I notice
you have done some homework' was the opening sentence of clinical
geneticist Mirjam van Beek in one of the consultations observed.[10] During
the counselling interview, the geneticist relies on the form to go through
the entire family history with the client and draw up a pedigree. If there
are too many hiatuses, the client is again encouraged to track down the
missing information. Only after the picture of the family history is maximally complete, does the geneticist starts addressing the client's own
health risks and the option of DNA diagnostics.

From the very beginning, then, the counselling trajectory requires the
client's active input in gathering data on her family history. This specific
role is a direct effect of uncertainties the clinical geneticist has about the
usefulness of DNA diagnostics, because by definition this technique does
not provide informative results. Furthermore, DNA diagnostics is too
expensive to apply to everyone who asks for it. To find out whether it is
responsible to deploy this new technology, the clinical geneticist needs to
have at least some indication of the likelihood of hereditary breast/ovarian
cancer and thereby the well-tried and conventional method of pedigree
analysis is the best available tool.

During this phase of the diagnostic process, the geneticist and the client
fall back on each other to satisfy various needs they each have. The client
needs the geneticist to gain more certainty about her health risk, but the
geneticist needs the client for mobilizing information that helps reduce
uncertainty about the usefulness of DNA diagnostics. However, this process
inevitably gives rise to new uncertainties. For instance, the role attributed
to the client presupposes that there are still family members who can be
reached and who can help to reconstruct the family history; that the
client is able and willing to contact these relatives; and that subsequently,
they are able and willing in their turn to tell about illnesses and causes of
death in the family. There can be a hitch on each of these counts.

After some hesitation Fiona, for instance, decided not to contact her
aunt and some cousins to get information on the anamnesis and cause

of death of her uncle, even though she was asked to do so in the first counselling interview.

> At first I did not mail it [the filled-out pedigree form] because I still wanted to write a letter to my cousin. I ultimately decided not to because I had great trouble finding the right tone. And when at one point my mother said, 'well, but he really never had cancer', I thought: I am not going to mess things up, for I really do not know how it will fall ... I mean, I haven't seen them in years; they might as well have just contracted some disease themselves.
>
> (Second interview: Fiona).

It turns out to be difficult, in this case, to contact relatives one has not seen for years, especially because illness and death are at issue.

In other cases, approaching relatives may not be an option on account of family feuds. For instance, when putting together Anna's pedigree, it is not clear whether her grandma has had cervical or ovarian cancer. The question matters because in the latter case the pedigree pattern suggests hereditary breast/ovarian cancer much more strongly.

> *Clinical geneticist Van Beek*: I think ... we have the opportunity to find it out because grandma is still alive; it would be well to gather these medical data, so as to know for sure that we ...
> *Anna*: It is difficult for us; we are no longer in touch with her.
> *Van Beek*: Not with grandma?
> *Anna*: No ...
> ...
> *Van Beek*: Who has contact with grandma and who doesn't?
> *Anna*: 'Well, only she has. [points to step-aunt in pedigree]
> *Van Beek*: Asking her is our only option, then, isn't it? It's complicated.
> ...
> *Anna*: 'I don't know whether she is going to participate; she is quite odd ...
> *Van Beek*: Whether grandma wants it. Yes; and if she does, she also has to consent ...
> *Anna*: Of course ...
> *Van Beek*: ... to hereditary testing. Yes, this is the issue you have to weigh.
>
> (First counselling interview: Anna)

It appears quite difficult for Anna to find out more about the details of grandma's illness because she is not in touch with grandma, nor is Anna's mother. What is more, the reason for their break-up proves to be tied closely to Anna's mother's breast cancer:

> *Anna*: Her mother [grandma] hasn't talked to her since the breast surgery [sneering laugh]. She [grandma] did not even bother to visit her.
> (First counselling interview: Anna)

It can be difficult, then, to get a full picture of the history of a particular disease in a family. The clinical geneticist has to rely on the client, who in turn is in part dependent on the quality of family relationships. At the end of the mapping process, a pedigree may still have many blanks.

Patient records from ill and/or deceased relatives are potentially useful in bringing down the number of uncertainties in a pedigree. Physicians can request these records, but here again the client has a specific role to perform: the family member involved or a direct surviving relative has to allow access to the record, and the client is the one to ask for their consent. At this stage too, however, the need for historical information can be too high a hurdle. The client should be capable and willing to ask her relatives whether the geneticist may inspect their record, and these relatives have to be able and willing to consent. Even if they do, it may turn out that the record has been destroyed. In the Netherlands, this is common in the case of older records; since the implementation of the Law on the Medical Treatment Contract (in Dutch abbreviated as the 'WGBO') in 1995, records have to be destroyed ten years after a person's death.[11] If access is possible, the geneticist should trace whether past diagnoses were indeed correct. A common problem in this respect is that evidence of ovarian cancer tends to show up late and is not diagnosed as such, but attributed to the organs that have metastases.

In most cases, the geneticist and the client have to try and produce a sufficiently informative family story in tandem. The geneticist needs the client to gain information about the family and to be able to request and inspect relevant records of family members, whereas the client needs the geneticist to be able to approach relatives in the right way and with the proper information. The success of this collaboration first and foremost depends on the client's willingness to take on that role. In part because most women do not anticipate such an active role, many have serious doubts:

> *Claire*: First, I had expected it to be very black-and-white, like: you want to have yourself tested, just raise your arm and we take a

blood sample, and in a few weeks you get the result. Well, it all proved to be quite different; it is time-consuming … You really need your entire family to trace that defect … You have to pay attention to who in the family has had the disease and at what age … It shows that you truly need their permission to inspect those records … This was a letdown; I just thought it would only involve me.

(Second interview: Claire)

Even when a client is prepared to collect information and ask for permission (and most clients are), for various reasons the fate of some relatives may remain unclear. A family's medical history is rarely known completely, which means that hereditary breast/ovarian cancer as a diagnosis has to be articulated on the basis of a limited set of data whose reliability is uncertain.

In this respect the distinction between hereditary, familial and sporadic breast/ovarian cancer as diagnoses is hardly productive, as geneticist Van Beek explained:

Because we cannot put it under the microscope and say 'this is hereditary and this is not', we use a definition and say: when three closely related people in several generations have had breast cancer, whereby one had it under the age of 50, we call it hereditary. This is not to say that when it fails to meet this definition it is never hereditary.

(First counselling interview: Anna)

These definitions will only advance a specific diagnostic process when enough family history data are available. This suggests that making a diagnosis involves a complex process, whereby geneticists have to make the most out of the available family data and the prevailing definitions.

This early stage of the counselling interview is mostly concluded with a diagnosis of the family history in terms of probability:

Arends: 'Well, all in all … there is much breast cancer occurrence at a fairly young age, as you will have gathered … This means that there is a substantial chance of it being a hereditary matter.

(First counselling interview: Doris)

The geneticist usually frames the message in a phrase like, 'in your family there is higher incidence of breast cancer than you would expect based on the average' – a message, indeed, that basically confirms what the client already knew. After all, this is why she contacted the clinic and got an appointment to begin with.

'Bothering others': DNA testing

If a probable family pattern of hereditary breast/ovarian cancer is established, the clinical geneticist tries to determine the client's risk of contracting breast or ovarian cancer. Such personal risk determination implies an estimation of the individual chances based on the pedigree and the epidemiological data.

It seems obvious that at this stage of the counselling process, geneticists bring in the relatively new DNA tests as the pre-eminent tool for making a risk assessment based on DNA features. However, healthy clients who come to the clinic as the first member of their family are rarely offered DNA diagnostics immediately. That geneticists tend to exercise restraint in suggesting this test to healthy women is related, as said, to its cost in combination with the slight chance that it will supply relevant information.[12] After all, the mutations found so far at BRCA1 and BRCA2 only explain a limited number of all cases of hereditary breast/ovarian cancer. For this reason geneticists would rather first test the blood of a family member who has actually contracted the disease. If this (ex-) patient carries a mutation, it is useful to test the healthy client for this same mutation; but if nothing is found, the advice usually is to refrain from testing healthy relatives.[13]

At this second stage of the diagnostic trajectory, the client is again attributed a specific role and responsibility because of the technology's limitations: she has to approach relevant family members (mother, sister, sometimes an aunt) and ask them if they are willing to have their blood tested for DNA traits. This assumes that these relatives can be approached and are willing to cooperate and donate a blood sample. Yet the first and crucial condition is the client's own willingness to confront her relatives with this request.

Many women consider their having to approach family members to be the most unexpected and difficult aspect of the entire counselling trajectory. One of them is Eva:

> What I remember of it, and this, I feel, is in fact the main point of this whole thing, is that to find out what potentially is in store for me has to be established via others, such as via my aunt and via my mother. This was something I had much trouble with, and still do, in fact. That you have to bother others with something that can be important to you and, later on, potentially, to your children ... was something ... I

hadn't thought of. This means that, to make it more clear to them and to just get a sense of 'what we should potentially be looking for', this really has to go via my aunt and my mother and that I enter into the picture only after them, while I was the one to initiate the whole thing – this I still find very difficult.

(Second interview: Eva)

Thus the practice of DNA diagnostics turns a seemingly straightforward question about individual health into an elaborate family affair. Even those who claimed to be aware of this from the start were unpleasantly surprised by the need for their own active involvement in the diagnostic trajectory. In the information brochures, this is addressed only implicitly:

A DNA test can only take place after pedigree analysis indicates the hereditary basis of breast and ovarian cancer. One will start, potentially, with a DNA analysis of an individual who already has or has had breast cancer or ovarian cancer.

(Dutch Cancer Society 2002).

Together you discuss the options for additional genetic testing (DNA diagnostics) and whether your relatives should also be involved. (When DNA diagnostics is useful for specific relatives, this can be performed only after individual interviews with them.)

(Polyclinic Clinical Genetics 2000).

Readers of these brochures may well gather from this that the cooperation of relatives can be necessary in DNA diagnostics, but nowhere is it indicated how one should ask for that cooperation, let alone what role the client herself thereby has to play. This role proved hard on many clients because they resented bothering others with their own need to learn more about their health risk.

Most geneticists know that it can be difficult for their client to discuss a request for DNA diagnostics. In the counselling interviews, they offer suggestions on approaching relatives and on reasons that may persuade them to cooperate. An oft-suggested strategy is to request their cooperation not right away but step by step, so as to allow relatives time to think about it:

Geneticist Arends: We do it by taking small steps anyway, mostly that is ... So the first step, which to her [Eva's aunt] is certainly somewhat easier, is to ask for permission to retrieve her medical data. I hand

you a form that she can sign and then we can retrieve her data ...
Next, you could ask her whether she would want to have DNA testing performed at all. And after we have the data, I will contact you
again and we can discuss here how things will proceed ... Thus she
will have time to think about the DNA testing.

(First counselling interview: Eva).

Mostly, the geneticist will offer a client periodic screening of breasts and/or
ovaries, so that during the waiting period she will not face extra worries
about her health.

Aside from strategies for approaching family members, geneticists also
suggest arguments clients might use to persuade relatives. One is that
the relative involved is or has been ill already, and that therefore the significance of the DNA diagnostics result will be much less to her than to
the (healthy) client. After all, it has few practical implications that this
relative, apart from being a (former) patient, may also prove to be a
mutation carrier. Sometimes this reasoning is reversed: it is indicated to
this relative that the result may be important to her because if she proves
to be a carrier, the remaining breast(s) will be monitored more carefully
while her ovaries will be periodically checked as well.

Given the importance in DNA diagnostics of information and blood
from relatives, the diagnostic process threatens to come to a halt if there
are no more (former) patients in the family who might be tested, or if a
client makes it clear that she does not wish, or is unable, to ask relatives
for their cooperation. In that case the geneticist will normally propose
starting with periodic monitoring of breasts and ovaries anyhow. In the
absence of (living) family members, it is sometimes proposed testing the
healthy client's DNA after all, and not just hers but that of all first-degree
relatives of the deceased family members with breast/ovarian cancer. If
by comparing the DNA of these first-degree relatives one traces no mutations, however, the result has a low predictive value:

> *Van Beek*: So ... there is now less of a possibility of detecting that
> defect in you. If you want DNA testing, one option is to test several
> persons. For you have a 50 per cent chance to carry that defect, but
> your brother also has a 50 per cent chance, and also the children
> of your late sister ... So what you could do is a DNA test of all four,
> because a defect might be found in you, for example, and not in
> them. And then *they* have a good result ... thanks to the bad one
> in your case. They will have certainty because it is found in you.
> But it may be the other way around as well ... So it is much more

complicated to do it now – and quite emotional, isn't it – because you need each other, and you depend on the result of the others. This makes it so hard right now ... If we do not find it in any of you ... then basically we still know nothing. We still do not know if a defect was present in your sister and your other brother. So you need to decide again: we may not have found anything, but isn't there still an equally good reason for monitoring? I think there is, because we have not yet excluded that there is no defect. DNA diagnostics, after all, does not allow us to detect all defects. If we fail to find a defect in your DNA, then it may still be there, given that we lack the proper technique for finding it. Only if it is found, for instance in your brother, while you don't have it, will we know how things stand. Then you do not have to be monitored any longer.

(First counselling interview: Irene)

Once again, the geneticist's diagnostic options depend on the client's collaboration. The client herself may be tested directly, but the number of relatives who are persuaded by the client to have themselves tested as well will determine the outcome's significance.

DNA diagnostics for breast/ovarian cancer basically allows geneticists to make claims about a health risk of an individual healthy woman. This technology seems a giant leap forward in respect to the old pedigree analysis. But this leap is only possible when clients first find one or more family members who are both ill and willing to have themselves tested. For this reason it is too simple to speak of 'the breast cancer test'. It is not just a matter of whether a woman is willing to hand in a few blood samples for a prognosis of her own health condition, but also a matter of her willingness to ask relatives for their cooperation.

'What is wisdom?' Deciding what to do

The counselling on genetic testing encourages clients to think about whether they actually want to know that they have an increased risk, and if so, which preventive measures they would want to take. In the light of the course of affairs in predictive DNA diagnostics for breast/ovarian cancer, however, this question is quite premature. Most interviewees did not so much struggle with whether or not they wanted to know what their health risk was, but whether the effort needed to find out was worth it. If DNA diagnostics requires much to be turned upside down, it may indeed be preferable to refrain from it and opt directly for regular monitoring. Therefore the issue boils down to: do I want to go on with the trajectory of DNA diagnostics *in this particular manner?* To afford a client time to think about this, the genetic counsellor and the client may decide in favour of periodic monitoring straightaway at their first meeting. This implies that the diagnostic trajectory and the decision process reflecting its specific character go hand in hand.

In contrast to the assumption made about the client's autonomy, moreover, women prove in actuality not to take such decisions alone. During pedigree analysis and the making of a preliminary diagnosis, as well as during discussions on what to do, client and geneticist work towards proper decisions together. In the consultations I observed, the client and the geneticist each brought in sources and performed activities aimed at potentially reducing uncertainties. As long as is needed they tinker, as it were, with the available sources so as to reach (preliminary) closure of the diagnostic process.

Such closure can take on many different forms, depending on the specific sources that geneticist and client have at their disposal. Rather than striving for absolute certainty on the client's health risk, the process is usually concluded when client and geneticist agree that the remaining uncertainties – in terms of their nature as well as their quantity – are acceptable and cannot in practice be further reduced. Thus a balance is constructed between the available options for extending the level of certainty on the one hand, and the effort needed to achieve it on the other. This balance can be different for each client, not only because clients have divergent values or views on the good life, but also because they have divergent possibilities for mobilizing sources.[14]

For example, in her first counselling interview Claire was told that her mother and/or aunt first needed to cooperate in DNA testing before testing her would be useful. Claire indicated that she did not dare to even approach her aunt because this woman denied that she had breast cancer. Claire did want to try to persuade her mother to cooperate, though, even if she knew that the chances were slight. Her mother, Claire felt, was happy to be cured (for the last three years) and wanted to leave her illness behind. This proved to be the case:

> Of course I first talked to my mother and she said: 'What are you letting yourself in for?' and: 'I don't even want to know about it.' And my father of course interfered as well: 'Leave your mother alone; it is hard enough for her already. You give your mother such feelings of guilt, as in "you brought us into this world ill, or such." ... You shouldn't do that; it really worries your mother, and perhaps it is all for nothing anyway.'
>
> (Second interview: Claire)

That the geneticist made the test less of a burden by mailing all the materials and blood test forms for Claire's mother to take to the nearest clinic was to no avail. Claire subsequently discussed the legitimacy of her mother's feelings of guilt and the importance of prevention in general:

> You can't help it being born this way, as a mother, can you? But as to not cooperating in testing, I said: 'There is actually something you can do now to turn the tide ... You also have a smoke detector in your home, and that sort of thing ... You can try to prevent certain things from happening.' But this argument hardly struck a chord with them.
> (Second interview: Claire)

Her mother, however, put her faith in the possibilities for early detection and treatment. As Claire pointed out:

> My mother said: 'Well, you see, once it is clear you have it, there is still a lot you can do, but now at least you are still ignorant. Once you know it, it is early enough to do something, just look at me ... They are so advanced these days, as long as you have yourself screened each time you're invited, they'll find nothing.'
> (Second interview: Claire)

When later on she tried to raise the issue once more ('informally, when doing the dishes') and her mother responded saying, 'Stop raising it, I don't like it', Claire decided not to pursue the issue anymore:

> Well, I am not going to raise the issue all the time, that much is clear to me ... I was thinking: Is there no possibility at all to find out more on my own? I also explored this option ... There was a programme about it on TV that I taped and watched when I had time for it ... But at one point I thought, well, okay, if I cannot do it alone I should merely stick to extra monitoring and that sort of thing.
> (Second interview: Claire)

She contacted the policlinic (several months after our first interview) to tell them that she had not managed to get cooperation from family members, after which the geneticist organized her referral for periodic breast monitoring in the hospital. Her conclusion reveals her resignation: 'I have done all I could, and if it doesn't work out, it doesn't work out.'

In another case, Janneke finally decides to refrain from DNA diagnostics as well and opt instead for periodic monitoring. While in Claire's case the bottleneck is lack of family cooperation, Janneke's decision not to continue is based on several considerations: lack of family cooperation, the limited reliability of the diagnostics, her own mental state and the risks or implications of potential preventive measures. Her case reveals the extent to which heterogeneous considerations need to be coordinated.

Janneke has two sisters who have both had breast cancer; one of them has two children, while the other sister and Janneke are childless. After the second sister contracted breast cancer, Janneke decided to go to a hospital, where a resident suggested to her that heredity examination would be a sensible thing. Both her sisters agreed but asked Janneke to find out more about this hereditary aspect. In her first counselling interview, however, the clinical geneticist told her that DNA diagnostics still had to start with her sisters. Subsequently the consultation mainly addressed what DNA testing could disclose, about Janneke as well as her sisters. As Janneke explained:

> What matters to me is: how do I see it. Look, now I have that 35 per cent chance, and I can say: Okay, I go see a surgeon and ... I can do the blood [test] for my ovaries and I can go in to have an X-ray once a year. There is really not any more I can do, not even if I know [that I am carrier] ... But what I win by not doing it [DNA testing] is that I just don't know it. Because it is some sort of roulette, isn't it? It is either 10 per cent or 80 per cent, and when you don't know it, you are at 35 per cent. And then there is the choice ... in terms of making yourself nervous, I mean. But to play it safe, to really be sure that you ... have an 80 per cent chance ... is like really wanting to be sure. But you don't know whether you will start worrying if indeed it is going to be 80. I am slightly psychosomatic; I did have various neurotic complaints ... so it does not come in handy with my sort of personality to add another anxiety level raising factor ... I do not have to know it, but I do all I can to prevent it. This, I think, is my choice, but most important in this case is what my sisters want. That, I believe, would be the only reason to opt for it [DNA testing] after all.
>
> (First counselling interview: Janneke)

She does worry about the potential risks of a periodic mammography:

> *Janneke*: Are X-rays really bad? The purists who say that X-rays are even worse for you, are they right? Once a year?
> *Arends*: I don't think so ... When you really have an increased risk of breast cancer, it overshadows [the harm done by] X-rays.
>
> (First counselling interview: Janneke)

In the second interview she reports that she talked extensively to her sisters and cousin, and eventually they all decided they could live just as well without DNA testing.

> Janneke reads aloud to the interviewer what she wrote to the geneticist: 'It seems that no one in the family really wants this test or finds it necessary. Annelies [oldest sister] only wants it because of her family; Cynthia, her daughter, doesn't want it because of uncertainties about insurance technicalities; my sister Theresa is not truly interested and simply assumes that it is not genetic ... nor am I anticipating such a test, even though I do want to focus intensively on preventive monitoring.' To the interviewer: 'So I finally thought that it will take up a lot of time, and ultimately it won't bring me anything, probably only anxiety. So I think: I'd better spend that energy and time on the [preventive] examination.'
>
> (Second interview: Janneke)

Together with the geneticist, and after consulting with her sisters, she arrived at the conclusion that for the time being, she was satisfied with a referral for periodic breast monitoring.

Janneke's consultation with the geneticist also triggered new uncertainties, however. It caused her to start wondering about whether she can continue to take the phyto-estrogens for menopausal complaints with her increased risk of breast cancer. She tackled this new uncertainty on her own. Because the surgeon who performed the breast monitoring could not tell her anything about this drug, she asked for information from the drug company, the association of pharmacists, a Chinese centre for herbal medicine, a naturopath, and a menopause consultant. The little information she managed to extract from these sources she planned to discuss at her next consultation with the surgeon, so as to learn whether the use of a soy extract can be harmful. Ironically, this could be a reason for having a DNA test after all.

> If finally they recommend I take no medication at all – be it regular, herbal or what not – I should perhaps participate in gene testing after all so as to learn if I really belong to the risk group ... This is, I believe, the reason why I would perhaps participate after all. Look, when you really know you have it, the risk is too large and you might as well have one hot flash after another and feel bad and unhappy about it. But, no, I still do not know the answer yet; and this means I go on mulling it over ...'
>
> (Second interview: Janneke)

In a way, Janneke, after the completion of the counselling trajectory, simply continued to obtain information on her risks. It suggests that the balance constructed in the genetic counselling trajectory is always a tentative balance, which in the case of new information or changed circumstances has to be actively constructed anew.

The validity of this observation also shows from the case of Irene. Right from her first counselling interview, when the DNA test is performed as well, she is referred to the surgeon and the gynaecologist, initially for periodic monitoring. Depending on the result of the DNA test, the geneticist explains, it is possible to discuss preventive breast and/or ovarian surgery with these specialists. Irene is adamant about preventive breast surgery being no option. About removing her ovaries she is less outspoken though, but the geneticist tells her that the technology used for checking ovaries is quite new and that they have little experience with it:

> *Irene*: And with ovarian cancer it is harder to see ... or discover than with breasts, true or not?
>
> *Van Beek*: 'Yes, at least you cannot check them on your own, because they are nestled deep down in your abdomen or, actually, low in the small pelvis ... This means you need to see the gynaecologist each year. He does internal examinations ... an internal ultrasound scan of the ovaries and a blood sample. In this sample, you look for the proteins. A single protein reading contains no information, but when it proves to go up, it can mean that an infection or cancer is developing in the ovaries. So this examination is very important, but it is also quite new. It is not that we have years of experience with it, as in breast monitoring ... and know what it results in. We believe that it is reasonably reliable, but we also know that it is not always reliable, that sometimes ... something may still be sitting there that you fail to detect.
>
> *Irene*: As good?
>
> *Van Beek*: Yes. So that applies to both. Through screening we have a much better chance, of course, ... because mostly you detect it on time ... but not always. So there is always the other option as well [preventive surgery]. These issues have to be discussed very well ... with the surgeon, with the gynaecologist; it is a matter of how things play out in your case, how well the breast images can be assessed and how well the tissue can be evaluated.
>
> (First counselling interview: Irene)

In this case too, then, the options to reduce an increased risk of breast/ovarian cancer generate new uncertainties. In the second interview, Irene reveals that she is still thinking about the preventive options:

> I heard [the surgeon] clearly tell me something like: 'Get them out, those ovaries.' When I said: 'Well, I am not so sure yet because removal of my ovaries will also increase my chance of cardiovascular disease, of a heart infarct, and these occur very frequently in my family as well.' So then I thought: 'Good God, what is wisdom in my case? I remove one thing and go down with another ... Dying of a heart infarct or gynaecological cancer ... which do you choose? I simply don't know ... I just don't know and I am not going to solve it.'
>
> (Second interview: Irene)

Fortunately, the gynaecologist appears to have new sources available for tackling these new concerns: she proposes to check whether Irene's body is perhaps already menopausal (at that point Irene is 50 years old).

> *Irene*: Well, there is a new perspective: examining whether or not I am menopausal because that very much plays a role as well. Not being menopausal and having your ovaries removed means quite a lot, doesn't it? This was also suggested by the gynaecologist and I was happy with it because I thought, okay, this is surely a thing to consider, isn't it? So this is what we are working on now.
>
> *Interviewer*: But what, exactly, is the reasoning?
>
> *Irene*: Well if you are already menopausal, or clearly getting there, and have your ovaries removed ... you suffer less from it ... And, of course, it is very true what the gynaecologist says: 'Look, if you are not yet menopausal, and we have those ovaries removed now, you'll soon need hormones to bridge the transition.' They are certainly not willing to give me those hormones, nor do I want them to because I am evidently in a risk group ... In such a situation, I can be sure I will ... contract cancer, you know?
>
> (Second interview: Irene)

The decision process, then, is not yet concluded nearly a year after the first counselling interview and four months after the DNA test result. That Irene proved not to be a mutation carrier complicates the issue instead of simplifying it. After all, a positive result (proven carrier) is the most certain result one can have. Its meaning is unambiguous, even though it involves a result in terms of chances. For a negative result, by contrast, multiple explanations are possible.

> *Irene*: No, no, I am not done yet ... And the gynaecologist takes it one step at a time of course, for she is required to do it step by step – so I couldn't tell you what it will be in the end! I don't know.
>
> *Interviewer*: Things may still go in any direction?
>
> *Irene*: Yes. And also because ... I clearly hear them say things like: you have a risk factor, but you do not have the gene, as far as we can tell. What to do next? So I am the one that has the hot pan! And this makes me wonder. I only consulted the experts to learn what wisdom is, but now I am the one who has to find out. What is wisdom in my situation? This is the hard part ... Well, at least I gained some breathing space, so as to determine whether or not I am menopausal right now.

(Second interview: Irene)

'Step by step', 'a new perspective', 'some breathing space' – in this specific case the diagnostic trajectory is still unfolding, and still is after the intervention of the gynaecologist and the surgeon. At this stage, as was true during the genetic counselling, the facts and readings provide the client with no certainty whatsoever – no basis for making decisions on how to proceed.

Irene's case clearly reveals that the physicians too have to determine as medical experts the uncertainty level that is acceptable to them. Is a pedigree analysis sufficient reason to refer someone for periodic mammography, or is such a basis too uncertain and should someone be a proven carrier first? Should periodic monitoring be offered to healthy people who cannot or do not want to mobilize family members for DNA testing, or should one first do comparative DNA testing of the eligible healthy family members? Do X-rays provide a sound enough basis of information, or should other diagnostics be employed as well? The eventual decision has to be in line with the views of both the client and the physician.

This is not a matter of the physician's repeating the patient's considerations all over again, as is seemingly implied by the concept of 'shared decision-making' or similar models.[15] The physician does not simply think along with the client about what would be best in her case, but also weighs which uncertainties she deems acceptable on medical-professional grounds.

Decision-making on diagnostics and prevention of hereditary breast cancer involves a process in which client and physician have to find common ground on heterogeneous considerations. Occasionally their perspectives may point in the same direction, as in the case of Irene and the gynaecologist, but they can also be at odds. While in counselling interviews the client is more or less forced to articulate and evaluate the (un)certainties she can live with, the clinical geneticist's professional ability to deal with uncertainty usually remains implicit.

'Another story about you': the added value of DNA diagnostics

The practice of DNA diagnostics for breast/ovarian cancer proves to be a trajectory in which uncertainties are constantly displaced and transformed. It involves a test-in-the making rather than a test, while its interrelated uncertainties are in part transferred to the client.

This conclusion is occasioned by two observations in particular. First, DNA diagnostics fails to provide a clear result in the vast majority of cases. Only in 25 per cent of the cases in which pedigree analysis seems to point to hereditary breast/ovarian cancer is a mutation at BRCA1 or 2 detected. In other words, the sensitivity and hence the predictive value of the current DNA testing is minimal, which was also reflected by the 11 trajectories observed. In all cases, pedigree analysis indicated possible hereditary breast cancer. Subsequently, DNA testing was applied to six participants or a family member (from five families), but a mutation was found only in one case (which had, in any case, been identified earlier in another member of this family).

This suggests that diagnoses in this field are often ambiguous. Although pedigree analysis (and the old techniques) may point to the possibility of hereditary breast/ovarian cancer, which means that a person who is a first-degree relative of a patient is a potential risk carrier, the actual DNA test may be negative. In such cases pedigree analysis is decisive, meaning that in the counselling interview the clinical geneticist will still emphasize that the test's negative result does not reduce the risk of breast/ovarian cancer and that, given the family picture, preventive measures are necessary. In the case of Irene, this double message was a hard buy:

I thought, okay, I don't have it, so we should all be happy, but no ... 'This or that is still possible, we're not there yet, and we don't know yet and ...' It makes me think: my god ... why did I go through it? When I'm still left with the sense that it may reveal itself ... anytime, especially taking into account my family history ... Everyone now says: 'Yes, but you have to view it as hereditary cancer' ... the family doctor, the surgeon, the gynaecologist ... And this was not clear enough to me, not clear enough in the preliminary examination.

(Second interview: Irene)

Although the limited predictive value was addressed in the information provided and in the counselling interviews observed, more explicit coverage of the limitations of current DNA testing would be useful.

Another observation, however, occasions the view that the technology's limitations are simply passed on to the clients. It turns out that DNA diagnostics makes little difference for the preventive measures that women ultimately take. In all observed trajectories, the outcome was the same: all women were referred for periodic breast monitoring, regardless of whether DNA diagnostics was applied (either to them or only to relatives) and whether a mutation was found at all. The only difference was that some women were also referred to a gynaecologist for monitoring of their ovaries. This mainly depended on whether ovarian cancer occurred in the family and on whether they had had breast cancer already.

One might comment that this study's sample is too small for making generalized claims on the choices made. Earlier studies, for instance, suggest that in the Netherlands a substantial number of women opt for preventive removal of breasts (Meijers-Heijboer, Verhoog et al., 2000; Meijers-Heijboer, Verhoog et al., 2001). Those studies, however, only apply to the choices of women in whom a mutation was detected, and this is but a small proportion of all women who come to the clinic. Moreover, the overall picture is distorted by the attention so far given in both public and professional media to mutation carriers and their predicament. The women who refrain from DNA testing or whose result is negative equally face tough challenges.

It is remarkable, to say the least, that the application of DNA testing seems to make so little difference to the measures that are eventually taken. This could well mean that DNA diagnostics does not so much result in specific preventive clinical implications, but in other self-images among the women who decided to do the test.[16] This much was suggested by Fiona, who at the time of the second interview had not yet decided if she wanted to have the DNA diagnostics performed:

They have to find it in her [mother] and next in me as well, or not in me; these are actually the only two clear situations. So yes, in this sense screening would already be enough, of my ovaries as well. I feel that then it would be possible for me to live with it, and leave it at that. Regardless ... that I am perhaps not a carrier at all, without me knowing it.

(Second interview: Fiona)

Karin, who did get a letter of referral for periodic monitoring, but who had not yet managed to motivate herself to make an appointment, put it even more succinctly.

By telling it this way I thought, yes, it could actually have been done without [the clinic]. What [it] adds ... is you get another story about your case from the one you would have had from your family physician.

(Second interview Karin)

Essentially, DNA diagnostics has little influence on women's preventive options.

Whether this will also be true in the future strongly depends on the genetic and medical specialists. Will they continue to consider it responsible to offer preventive options to all women with an indication of a higher familial risk? Or will they weigh DNA diagnostics more heavily than family history and pedigree analysis, and only offer preventive options to women with a mutation? One of the clinical geneticists questions offering periodic monitoring to all women:

I and a great many of my colleagues throughout the country are becoming ever more directive in this respect. Because we still believe it to be proper medical action to do DNA testing ... And this has to do not only with my sense that they are at risk themselves and perhaps may avoid it ... But also because I think that people perhaps wrongly make use of medical care. For instance, when people prove not to be a carrier of the mutation found in the family, I consider it medically incorrect that they have to have a mammography and see a surgeon each year. Not only because of the capacity problem in surgery, but also because of the X-ray radiation. People should not be exposed to even a little of it. And the same applies in the case of the ovaries. In the general population there is quite a low risk of ovarian cancer. If you offer someone yearly monitoring because you do not know if this person carries the mutation while you do know that she is potentially in a risk category based on the pedigree, they can say: I don't do a DNA test because I cannot handle it. And then we say: but make sure you have it checked regularly, meaning that each year they get gynaecological screening ... It makes me think: this is medically incorrect ...

because you still may be offering unnecessary intervention to people, with unnecessary narcosis and such. It involves unnecessary use of the medical capacity. And this is what the surgeons and gynaecologists increasingly tell us, especially when they know that there is a mutation in the family. They no longer want to take in people for screening if it can be tested.

(Interview: clinical geneticist Van Beek)

Van Beek's reasoning implies that physicians will perhaps push DNA diagnostics more, because they want to be sure that periodic monitoring and/or preventive surgery are not deployed redundantly.[17] This may involve a medical argument (not wanting to expose clients unnecessarily to health risks), but also a financial-economic argument. In the latter case, DNA diagnostics becomes an instrument for realizing efficient spending of collective means. But in order to gain such benefit in the medical-technical and the economic domains, the sensitivity of DNA diagnostics will have to increase significantly.[18] Today DNA diagnostics can fulfil this selection function only regarding families in which a mutation has already been identified.

Responsibility of clinical geneticists

Predictive DNA diagnostics for breast and ovarian cancer relies today on an all but ready-made 'test'. This form of diagnostics implies analysis of body material in the lab, but as such the 'test' does not produce useful information. DNA diagnostics is a comprehensive social practice that aside from DNA analyses comprises the mobilization of family histories, records, genetic knowledge and definitions and, last but not least, family members, whereby commonly more than one person needs to have their blood tested before something can be said about the risk of the individual client. Speaking of a 'breast cancer test' obscures how much work is needed to generate an informative result. This also means, though, that for the time being future scenarios in which large numbers of healthy people are subjected to this predictive diagnostics are hardly likely.[19] After all, at this point DNA diagnostics for breast/ovarian cancer comes with too many uncertainties and limitations, while it is also time-consuming and expensive.

To overcome the uncertainties and limitations of this genetic technology, the clinical geneticist and client have to muster many resources. Accordingly, the counselling trajectory can be conceptualized as a shared tinkering with divergent tools and means to fight a host of uncertainties.

The trajectory arrives at a (preliminary) conclusion when in some form or another closure is achieved. This closure does not so much involve certain or definite diagnosis, but a pragmatic balance between the possibilities for acquiring yet more knowledge on the one hand, and the burden associated with the effort needed on the other. Such balance is achieved when the nature and amount of remaining uncertainty is deemed acceptable by both client and physician. According to the ideal of autonomy in genetics, the distribution of roles between client and geneticist, in the construction of that balance, should be such that the geneticist supplies the relevant information and the client makes the moral choices. My argument has demonstrated, however, that this picture does not coincide with actual practice. Contrary to the suggestion of the autonomy ideal, in the diagnostic trajectory there is no neat distinction between facts and values, or between providing information and making decisions. Moreover, the decision process is highly pragmatic; decisions often depend more on what the client and the geneticist both deem 'feasible' or 'doable' than on what clients consider morally desirable. Therefore, what is usually presented as a highly personal moral choice between, for instance, knowing and not-knowing, involves language that is much too strong.[20]

This said, it should also be observed that the work and responsibilities involved in this pragmatic process are not equally distributed between geneticist and client. While the geneticist can do only little in terms of providing information, and finally does not have to decide on DNA testing and possible preventive measures, their female clients ultimately have to make the decisions as well as do the preliminary work. Notably the healthy client who goes to a clinic for clinical genetics as the first of her family is expected to turn much in her life and family upside down in order to allow for the subsequent successful application of DNA diagnostics. The specific predictive value of DNA testing depends in fact on the degree to which she is able or willing to mobilize family members. Clients, it turns out, consider this a heavy responsibility.

One might argue that the information needs to be much more explicit on this point. As already suggested, there is certainly much to improve in this respect. And yet this is an unsatisfactory solution because it puts the burden of an imperfect technology one-sidedly on the client. This is worrisome, all the more so because DNA diagnostics for breast/ovarian cancer in most cases provides no more certainty on someone's risk of contracting these diseases than the 'old' pedigree analysis. While the burden of DNA testing is substantial to women, the benefit is but slight: the preventive options are the same. It seems unjust that clients largely bear the burden of a test-in-the-making and its uncertainties, while gaining so little from it.

This injustice can be diminished if clinical geneticists take on more responsibility for the technology's shortcomings. As long as the predictive value of DNA testing is not improved, they should refrain from DNA diagnostics in healthy individuals, be it their client or their client's relatives. Precisely because DNA diagnostics implies more than just testing a single person, healthy persons who go to a clinical genetics clinic and ask for DNA diagnostics should be rejected if in their family no mutation has as yet been found. They are still eligible for care: one can offer them the same surgical and gynaecological monitoring that today most women who ask for it already receive, regardless of whether or not their DNA was tested.

Back to the patients

Predictive DNA diagnostics, however, does not have to be discarded as a specific medical service. The problem sketched above occurs in particular, as indicated, in predictive diagnostics for healthy people from families in which a mutation has not yet been identified. DNA diagnostics in *patients* (which does not so much predict a diagnosis but confirms it and makes it more precise) is much less problematic. This provides specific starting points for an alternative practice of predictive diagnostics, in which the responsibility for reducing the uncertainties in the technology would mainly be with the medical staff. Instead of approaching ill relatives via healthy ones, healthy family members might be approached via ill ones.

What would such practice look like? It might start in departments of surgery and gynaecology, where it could be analysed, possibly with help from a geneticist, whether heredity could be at play for each breast/ovarian cancer patient. One could even pursue this as part of an active policy. If the family history warrants consideration of a hereditary factor, one could subsequently offer the patient DNA testing. This test could then be done without involving relatives. If the test reveals that the patient is the carrier of a mutation, this information can be used in decisions on that patient's treatment and monitoring policy. In addition, one could ask the patient who carries the mutation to distribute a letter from the genetics clinic among all first-degree relatives (as already happens when a mutation is found in a healthy or ill family member). This letter informs them about the possibility of having their own DNA tested. Thus the choice whether or not to opt for testing remains in the hands of the person involved.

This reversal of the current practice of predictive diagnostics has two major advantages. First, one can offer a test to the healthy relatives involved that does in fact provide relevant information: because it is already known that there is a mutation in their family, it is possible to determine with quite a high level of certainty whether or not they are a carrier of that same mutation. Moreover, this method of approach is less drastic for both patients and their relatives. Although the patient still plays a role as intermediary in approaching relatives, a requirement which on account of privacy rules cannot simply be altered, in this scenario she does not have to approach those relatives because she herself wants more information about her own health risk. She has, in other words, no direct interest in the choice made by her relatives, which makes her task much less of a burden.

Does this proposal for changing the practice have any chance of being successful now that predictive diagnostics has for some years been offered to healthy individuals? It would imply after all that clinical geneticists retrace their steps and exercise restraint in offering this technology. My argument underscores that such moderation is warranted. The limited number of informative results unambiguously demonstrates that predictive DNA diagnostics for breast/ovarian cancer has as yet failed to move beyond its development stage. This particular diagnostic technique is a semi-finished product rather than a test. As long as this is the case, clients should be burdened as little as possible with this technology's imperfections.

Acknowledgements

The fieldwork for this chapter was done in the context of the project 'Problems of social cohesion and health politics in an era of predictive medicine', which was funded by the Netherlands Organization for Scientific Research (NWO) research programme on Social Cohesion. I would like to thank all interviewees, clients and physicians for their willingness to cooperate in this research project. I also want to thank the clinic staff member who was willing to be the first to ask clients for their cooperation.

4
Lifestyle, Genes and Cholesterol: New Struggles about Responsibility and Solidarity

Klasien Horstman

Geneticization?

Since the start of the Human Genome Project there has been much speculation on the impact of this mega-research project on medicine and
health care. Various authors have argued that growing attention on the
genetic factors in the emergence of diseases will lead to their geneticization: it would define diseases increasingly in terms of DNA (Ten Have, 2001).
Geneticization, according to the sociologist Lippman, means 'the ongoing
process by which priority is given to differences between individuals based
on their DNA codes, with most disorders, behaviours and physiological variation ... structured, at least in part, hereditary' (Lippman 1993, p. 178).
Several authors fear that the rise of this genetic perspective will result in a
devaluation of social or cultural approaches to health and disease. After all,
if genes are decisive in determining the boundary between pathology and
health, is there still any relevancy to social or cultural explanations for
health or disease? Will there be room left for individual meanings of health
and disease? The geneticization thesis implies the expectation that on
account of the increased usage of genetic tests, people are tied to their biological fate and that their freedom to fashion their own lives is seriously
threatened.

The geneticization thesis has met with criticism, however. This criticism primarily has a methodic character. According to Hedgecoe, the
concept of geneticization is much too speculative and monolithic for an
adequate analysis of the effects of the introduction of genetic technology
in health care (Hedgecoe 2001). This concept goes hand in hand with the
view that geneticization is a necessary and unavoidable effect of genetic
technology and does little justice to the *human effort* involved in the introduction of new medical techniques. The geneticization thesis would also

shed insufficient light on the influence of critical discussions about DNA technology as to how these techniques ultimately become embedded in society. When starting from the geneticization thesis, little is to be expected from public debate and reflection because most power is attributed to the technology itself. Based on this specific critique, Hedgecoe argues for investigating the effects of genetic technology, rather than simply assuming them. This will allow one to assess whether the claim implied by the geneticization thesis – the increasing definition of disease and health in terms of genes – will hold. He considers it the task of the social sciences to enrich the debate on genetic technology with empirically informed analyses.

In this chapter, which takes Hedgecoe's suggestion to heart, I trace the introduction of genetic technology in the area of cardiovascular diseases with the aim of making visible the human effort involved and to generate new normative questions and reflection, instead of creating fear.

During the past four decades, cardiovascular diseases have mainly been associated with the risks of a specifically Western lifestyle. Although recently these diseases[1] have also been linked specifically to the dietary and smoking habits prevalent among members of the lower social-economic classes, the risks of the Western lifestyle as such apply to the overall general population. In view of the traditional attention given to lifestyle risks in the prevention of cardiovascular diseases, the question arises what the new genetic-diagnostic techniques imply for our perspective on such diseases. Does the introduction of genetic techniques indeed result in a shift of attention from the risks of a particular lifestyle to the risks of specific genetic aberrations? If so, how does it work and how does it affect the control and responsibility that we attribute to individuals when it comes to their health?

To answer these questions I take a close look at one specific example, namely the introduction of a genetic approach to hypercholesterolaemia as developed in the Netherlands over the past 15 years. In this chapter, I analyse how this new approach relates to the already existing views and practices involving cholesterol. It will become clear that it takes a lot of effort to define a genetic risk group, but that the demarcation of this risk group is not accepted without resistance. In fact, the genetic high-risk approach clashes with existing approaches in which the general population's risks are centre-stage. As such, this struggle is not just about defining risk groups, but also about assigning responsibility for health or disease and legitimizing the spending of public funds. Because this struggle is far from over, this analysis merely provides a limited perspective on the issues involved. It becomes clear, however, what sort of controversies are in

store for us with the introduction of other DNA techniques in health care –
and what the stakes will be. But before addressing genetics, we first need to
go back in time and consider the emergence of the prevention of cardiovascular diseases.

From fate to risk

In the middle of the twentieth century, cardiovascular diseases were recognized as the main cause of disease and mortality in Western countries.
Some even referred to it as an epidemic. Despite much experimental medical research, there were still no adequate therapies within reach and in fact
no one considered this to be odd. After all, both professionals and laypeople had long viewed cardiovascular diseases as the expression of a natural
aging process. The famous Framingham study that started in 1948 would
change this view.[2] In the context of this epidemiological study, the so-called
risk-factors approach to cardiovascular diseases was developed, which
caused them to lose their taken-for-grantedness while prevention became
a valid option. If there was no therapy, one could structurally change disease and mortality patterns through the early detection of risks. Prevention
seemed to be the royal road toward public health. Following on from the
complaint-oriented approach to cardiovascular diseases, a preventive and
risk-oriented approach had entered the field.

With the introduction of thinking in terms of risks, however, it was not
immediately clear what the relevant risks were. The Framingham study
identified as many as 300 risk factors for cardiovascular diseases, while
other epidemiological-statistical studies also reported an array of risk
factors. As one commentator put it: '... raised blood pressure, cigarettes,
lack of physical exercise, obesity, "nervous stress", hyperlipaemia, diabetes,
heredity – this reminds one again of the parable of the blind who each in
turn had touched the elephant'(NTvG 1965, p. 2101). The necessity of
prevention through a focus on specific risks was, however, increasingly
acknowledged: 'Risk identification and modification is the only medical
approach now available for effective reduction in the burden of coronary
heart disease' (NTvG 1970, p. 347). Why did so many professionals advance
this risk-factor approach as a major promise in fighting cardiovascular
diseases?

This had to do, according to the historian Aronowitz, with the fact
that this approach provided a conceptual frame for the ambivalent
attitude among many physicians with respect to experimental medical-
scientific approaches of disease (Aronowitz 1998). Although a strictly
scientific approach to the body enjoyed great prestige, many clinicians

had second thoughts about the physiological reductionism and the disappearance of the individual patient in medicine. It was the vocabulary of risk-factors that made it possible, in Aronowitz's estimation, to move away from the reductionist and monocausal explanations for cardiovascular diseases that were favoured by those working in laboratories, while the quantitative empirical style of epidemiology could still be labelled as scientific. In other words, the risk-factors approach made it possible to conceptualize the contribution of individuals in generating cardiovascular diseases, which provided specific handles to clinicians in the consultation room ('mind your diet!') and also allowed for risks to be seen as natural facts that could be mapped objectively using scientific means.

From 1970, the risk-factors approach materialized in national and international public health policies aimed at pushing back cardiovascular diseases.[3] As stated, the Framingham study had listed 300 risk factors, but in the context of the conceptualization of cardiovascular diseases as diseases of affluence that struck the entire Western population, three risk factors crystallized as the starting point for national and international prevention programmes: smoking, bad diet and lack of exercise. These symbols of a Western lifestyle became the central focus of efforts to change the way citizens of Western countries were living. As such, this unhealthy lifestyle had long been regarded as a *collective* problem, and many therefore emphasized the responsibility of the government to alter this situation.[4]

Since 1990, however, we have been witnessing a turn in such thinking. Smoking, a fat diet and little exercise are regarded less as symbols of Western *culture* and increasingly either as signs of *individual* lack of willpower and self-control, or as an expression of the deliberate individual choice of a hedonist lifestyle. In short, instead of symbols of a lifestyle that we automatically adopted in our youth, that we share with many and that is stimulated by media, the economy and government, lifestyle risks began to be viewed as a matter of individual responsibility (Lupton 1995). However, this process of individualizing risks cannot obscure the fact that for decades much was expected from general lifestyle changes when it came to the prevention of cardiovascular diseases. Although everyone acknowledged that predisposition of course plays a role in the emergence of cardiovascular diseases, this was rarely a factor in the design of prevention programmes during the last three decades of the twentieth century. The question arises, then, what the introduction of molecular biology and genetic diagnostics has done in respect to the lifestyle paradigm and its associated conceptualization of risks, prevention and responsibility.

The promise of genetics

The progress of the Human Genome Project has also caused expectations to run high regarding the role of genetics in the prevention of cardiovascular diseases. As cardiologist R. Roberts puts it:

> Prevention will be the key to future successes, and the unraveling of human genes will catapult prevention as a major initiative for the 21st century. In the management of coronary heart disease, individuals will be identified in their teens so that treatment and prevention appropriate for their medical risk profile and life style can be properly individualized.
> (Roberts 2000)

If Roberts is very optimistic, others share his intuition that genetics will strongly change our approach to cardiovascular diseases in the near future (Brugada 2000; Ellsworth, Sholinsky et al. 1999). Similarly, many epidemiologists have embraced genetics. They view the individual genetic risk profile as a vehicle for rendering prevention more successful than ever. Their assumption is that knowledge about the risks for cardiovascular diseases, rheumatism, cancer et cetera will allow someone to lower their chance of contracting these diseases through medication or a specific lifestyle (Khoury, Burke and Thomson 2000; Nora, Berg and Hart 1991; Omenn 2000).

A common response to high expectations about a new field is scepticism: things won't come to all that. But even though boosters of new medical techniques are likely to paint a too rosy picture of the particular health benefits gained in the short run, the promotional effort on their part can still play quite a stimulating role in the formation of new research and user practices. After all, researchers, funding facilities and users will start acting on raised expectations and thus contribute to actually realizing them (Van Lente 1993). For instance, the science scholar J. H. Fujimura has demonstrated how the promise of molecular biology played a major role in creating a new *bandwagon* in cancer research: the promises basically challenged scientists with various theoretical and methodological baggage to jump on to this bandwagon, which indeed contributed to the success of a molecular-biological representation of cancer (Fujimura 1988). In short, even if promises are not realistic, they may still have effects and therefore we have no choice but to take such promises seriously. Unmistakably, genetic practices relating to cardiovascular diseases, in the laboratory as well as in society, are currently evolving rapidly.

The Dutch Heart Foundation, for example, has capitalized on the expectations articulated by Roberts and others. In 2001 its public fundraising campaign was geared to the heredity of cardiovascular diseases, and in its brochure *Heredity in cardiovascular diseases*, we can read:

> In recent years heredity has attracted much attention. This is the case because so much more has become known about the 'human genome', the basis of our hereditary features. Scientific researchers discover ever more genes that are responsible for the emergence of specific diseases. We can turn this knowledge to our advantage. Bit by bit, there is also more information on heredity in cardiovascular diseases. In this brochure the Dutch Heart Foundation has summed up this new information. Unfortunately we do not know everything yet.
>
> (Dutch Heart Foundation 2001, p. 3)

Despite the modesty conveyed in the brochure's introduction, the promise of genetics is emphasized once again on its final page:

> In the future more people will be able to learn whether they have a raised hereditary risk for cardiovascular disease. This offers the opportunity to take timely measures and limit or avoid adverse effects.
>
> (Dutch Heart Foundation 2001, p. 18)

And before the Dutch Heart Foundation informed the public of the great promise of molecular biology, in the early 1990s a nationwide programme had been launched aimed at the detection of genetic risks for *familial hypercholesterolaemia* (FH). This detection programme did not emerge in a vacuum, however, because attention to cholesterol levels was already part of an established practice.

Cholesterol in general practice

From the 1970s onward, in most Western countries a preventive practice with respect to cholesterol began to be developed. A high cholesterol level was defined as a risk for cardiovascular diseases because it hardens the arteries; instruments were introduced to measure cholesterol in the blood; standards for normal and pathological values were formulated and one began to make a distinction between good and bad cholesterol. In 1995 Schuurman concluded in her dissertation *Blood cholesterol: A public health perspective* that in recent decades the mortality of coronary heart diseases had gone down but that it was still high; that the cholesterol levels in the population were still high as well; and that cholesterol lowering in the general population continued to deserve support. As she claimed:

> This cholesterol lowering has to be achieved preferably by changes in lifestyle (a diet with a low level of saturated fat, many vegetables and fruits, no smoking, no obesity and regular physical exercise).
>
> (Schuurman 1995, p. 166)

Although company doctors also began to measure cholesterol, it was especially in general practice that cholesterol testing gained ground. Because GPs were increasingly confronted with requests for cholesterol measurements and new cholesterol lowering drugs had been introduced, a so-called 'cholesterol standard' was formulated. Such a standard is considered to be a part of the well-founded guidelines that count as 'best practice' for GP's medical practice. The first cholesterol standard was published in 1992, followed by a second, revised version in 1999. Given the fact that many GPs were involved in the preliminary process, we may assume that both standards adequately reflect how they thought – and think – about cholesterol.

The first standard from 1992 shows that GPs share Schuurman's focus on overall public health (Binsbergen, Brouwer et al. 1991). For instance, in the introduction to the standard it is argued:

> General health information campaigns may lead to a reduction of 5 per cent in the average serum cholesterol level in the population. If this reduction at the population level is the most effective, it does not belong to the GP's task.
>
> (Binsbergen, Brouwer et al. 1991, p. 551)

In the standard's explanatory commentary, the authors claim that the effect of the population strategy is twice as large as that of the selective strategy. The GP's task, then, is a supplementary one – mainly geared to people who, despite the public information campaign, still continue to have high cholesterol. Regarding the improvement in public health, the authors expect most will come from the population strategy. Moreover, they indicate that it is not a settled issue that both prevention strategies – the individual approach in relation to high risks and the general population strategy – are not necessarily complementary. More attention on high risks might well contribute to reduced motivation to change one's lifestyle among the entire population.

Second, the standard's authors explicitly claim that a high serum cholesterol level is not a disease, but one of the risk factors for coronary heart and vascular diseases. In their explanatory commentary, we read: 'The predictive value of raised serum cholesterol for developing CHZ is limited' (NHG 1991, p. 4). The importance of cholesterol, so the authors argue, depends on the specific cholesterol level and its interaction with other risk factors, while it is not possible to draw a clear boundary between normal and abnormal cholesterol. This is why the cut-off points that are mentioned in the standard are presented as pragmatic boundaries rather than absolute criteria. To illustrate this, the authors report on the different cut-off points in various countries.

Dutch GPs tend to exercise restraint when it comes to cholesterol measurement and treatment. In the standard it is argued that people's cholesterol varies and that a cholesterol reading needs to be repeated to produce a reliable result. They emphasize the complexities involved in the occurrence of cardiovascular diseases, the limited autonomous significance of cholesterol, and the lack of insight into the specific effect of many risk factors. They are aware, so the standard indicates, of the risk of over-diagnostics and over-treatment, of medicalization and of the cost of medication. In the context of this style of reasoning, the standard sets a restricted indication for cholesterol measurement (patients with cardiovascular diseases, with familial cardiovascular diseases – such as FH – in the family, with hypertension and diabetes mellitus); in cases of raised cholesterol levels dietary advice is provided first. Only if the serum cholesterol level does not go down is medication considered. The standard mentions familial hypercholesterolaemia as a major risk factor for cardiovascular diseases, but gives it no more attention than other major risk factors.

The 1999 revised version of the standard did not change much in terms of style (Thomas, van der Weijden et al. 1999). Again it emphasized that raised cholesterol is a poor predictor of cardiovascular diseases because most of these disorders occur in people with a normal or slightly raised cholesterol level. And once more the multifactorial character of cardiovascular diseases was stressed: 'The various risk factors reinforce each other and need to be viewed in their mutual interdependency' (Thomas, van der Weijden et al. 1999, p. 407). The major change in this standard in relation to the previous one is that the expectations with respect to a healthy diet have been adjusted. The indication for cholesterol measurement is more detailed and those who prove to have raised cholesterol are prescribed medication immediately. This shift from nutrition to medication is a reaction to the wider availability of cholesterol reducers, but also to the fact that nutritional advice often runs up against the unruliness of patients' everyday life. The shift does not mean, however, that GPs discarded their earlier views altogether. In the explanation of this shift, for instance, they qualify expectations regarding medication, taking note of the fact that while its effects are calculated based on a 25-year period, the effects were actually studied in practice only for a period of six years. Still, the authors of the revised standard emphasize that stopping smoking is of great importance and can even make medicinal treatment superfluous. People with FH are mentioned and treated as high risk, but empirical data on their risk 'are scarce' (Thomas, van der Weijden et al. 1999, p. 414).

The paradigmatic style of GPs reflected in the cholesterol standards clashed with the perspective of some internists involved in cholesterol research and working with a very different patient population from GPs. For instance, the authors of the 1991 standard observe that countries with lower cholesterol levels and lower incidence of cardiovascular disease had no higher life expectancy and it was thus questionable whether cholesterol reduction was always a good thing:

> It has not been ruled out that a higher non-cardiovascular mortality goes together with lower cholesterol values. This would imply that 'the lower the cholesterol, the better' is only valid in part. It is not clear, then, what a safe lower limit is for the serum total cholesterol level.
>
> (NHG 1999, p. 3)

In an article in the Dutch *Journal of Medicine*, Erkelens, an internist, dismissed the above reasoning – which questions the assumption that cholesterol lowering is always beneficial to a person's health – as a myth (Erkelens 1993). However, in a reaction Van der Weijden, one of the authors of the standard, argued that such a view could not yet be corroborated because insufficient empirical data was available and therefore the issue of whether lowering a raised cholesterol level was always safe still stood (Van der Weijden 1993, 1420). Similarly, a discussion between the internist Kastelein and the standard's working group about the nutritional advice revealed that GPs and specialists have different approaches. Kastelein argued that the standard's attention on nutritional advice suggested that the standard was aimed at a large group of people with moderately raised cholesterol and that consequently, FH patients ran the risk of being wrongly treated:

> Your standard creates the possibility that a 30-year-old man with FH who is a smoker and displays no 'symptoms of a familial

hypercholesterolaemia' but who has a cholesterol level of 9.5 mmol/l is fobbed off with a diet for years.

(Kastelein 1991)

In response to this protest the standard was indeed adjusted.

These differences of opinion illustrate well the divergent approaches of GPs and internists. The former expect most from the population strategy; they are not inclined to magnify or take the reliability and predictability of the cholesterol test as absolute, and they continue to point to the effects of the combination of risk factors while in terms of intervention aside from medication, they expect much from lifestyle change. In contrast, specialists are more geared to a select risk population and emphasize the significance of cholesterol as an autonomous risk factor and of medication as intervention. In Aronowitz's terms, GPs represent the risk-factor approach in preventing cardiovascular diseases, while specialists are more inclined toward a reductionist approach. In the controversy about the value of detecting FH, we will see the tension between these approaches come to the surface in a much more charged form.

The GPs' approach implies that they do take FH seriously as a risk factor for cardiovascular diseases, but that they make no distinction between this high risk and other high risks. Evidently a substantial effort is still needed to organize support for the detection in terms of genetic risks only.

The detection of FH: the construction of difference

According to politicians and ethicists, a nationwide detection programme for specific risks has to meet a number of conditions. First, it has to be proven that the diagnostic techniques involved are adequate and that detection in a medical sense is useful. Furthermore, there should be no large social and ethical objections. The technology-philosophical view implied in this reasoning assumes that both the development and application of technology are clearly distinct stages. It assumes that a diagnostic test is fully developed when it is first used on a large scale and that decisions on large-scale detection are based on facts about the value of the test rather than on assumptions or postulates. The value of a test must be proven before its application. The scientific work, so to speak, has to be completed prior to the technology's large-scale application. Various scholars in science and technology studies have pointed out, however, that this particular way of representing things is debatable and that, for instance, the application phase can also be viewed as a continuation of the experiment in another context (Latour 1987, 1988; Bijker and Law 1992).

Contrary to the idea that development and application, or scientific practices and social practices, are distinct domains, various authors have shown that they evolve simultaneously and in mutual interdependence. For instance, the use of a specific diagnostic test is a necessary condition for the development of that test. Laboratory and clinic depend on each other, and this certainly applies to genetic knowledge and technology. To improve DNA diagnostics, reveal new mutations and determine their significance, it is imperative to gain insight into the clinical picture of families. During that process, the boundaries between a test-in-the-making and a ready-made test often become blurred. Yet in order to be able to develop the test further, it is often presented as ready-made because this stimulates its actual utilization and further development. In other words, claims about utilization often function as facts and this is one of the mechanisms by which claims can ultimately become facts.[5] In order to let a technology live up to its promise, its uncertainties are under-represented while claims that favour it are frequently put forward with more certainty than is warranted. We also see this mechanism at work in the way in which the special meaning of genetic risks involving FH was discussed in the early years of the detection programme.

In the late 1980s, researchers from the Amsterdam Medical Centre (AMC) had already initiated study of hypercholesterolaemia's genetic background. Inspired by work in other countries involving families with pathological lipid patterns, serious cardiovascular diseases and high mortality at a young age, they set up a lipids clinic for examining patients and their families. When in 1990 the literature reported the first mutation in the Netherlands and it became clear which gene was involved in high cholesterol levels, the detection of families and the study of new mutations were intensified. That in 1989 a new medication called 'statines' had been introduced for lowering cholesterol, seemed a happy coincidence. Now, at last, it was possible to link the detection of families to a therapeutic perspective: there was something to offer to people! The first studies and detection activities were funded by the pharmaceutical industry, but in 1994 a specific foundation, the StOEH (Hereditary Hypercholesterolaemia Detection Foundation), was set up with the objective of detecting everyone with FH in the Netherlands. Its establishment implied that a private research initiative was being transformed into a public health care initiative funded by public money.

Although research and detection were now formally dissociated, the research motivation remained linked to detection. In order to continue the research of other mutations and their specific function, the AMC lab did not confine itself only to the analysis of DNA taken from detected families but, in collaboration with StOEH, also went on to study the connection between genotype and phenotype. This effort perfectly exemplified the co-evolution of scientific and social practice. A genetic-diagnostic test for FH was being developed and the organization of the detection of yet other families – still more DNA and more pedigrees – fulfilled a crucial function in the further development of this test and of the insight in the relationship between mutations, raised cholesterol and cardiovascular diseases.

In light of the uncertainties initially linked to DNA diagnostics for FH, it is remarkable that this detection programme became integrated in Dutch health care so quickly. In this chapter, though, I do not address the political process concerning the detection programme. What I would like to demonstrate here is how, in the underlying reasoning for the detection programme, *difference* is constructed between, on the one hand, genetic risks for raised cholesterol and cardiovascular diseases that affect a limited group and, on the other hand, the raised risks of high cholesterol and cardiovascular diseases caused by an unhealthy lifestyle and affecting the entire Dutch population. Although it is explicitly argued that a genetic approach and a lifestyle approach can go together very well, the particular usefulness of genetic detection is argued in competition with other prevention strategies. Thereby geneticization is at stake to the extent that a specific form of cardiovascular disease is presented as caused by a specific gene.

The various policy plans, annual reports and brochures of the StOEH and the information materials issued by the patient group Bloedlink allow us to trace a chain of arguments that contribute to the construction of FH as a genetic disorder.[6] First, FH is discussed in terms of a commonly occurring genetic disorder. In contrast to the notion that genetic diseases occur infrequently and that only a few families are smitten by them – up to this point DNA diagnostics was mainly geared to genetic disorders that indeed were rare, such as Huntington's disease and muscular dystrophy (Nelis 1998) – FH is presented as a fairly *often occurring* disease, one that might strike any one of us:

> Familial Hypercholesterolaemia is the most frequently occurring congenital metabolic disease. This disorder occurs so often that many Dutch people are likely to have someone in their circle of friends and acquaintances who suffers from this problem.
>
> (Kastelein and Defesche 1995, p. 3)

This image of FH as a 'normal' disease in terms of its prevalence is supported by statistical information. According to estimates, one in every 500 people will have FH. Although it may be a hereditary affliction, the suggestion is that all of us are somehow connected to it.

Subsequently, this normalization of FH is halted by indicating that FH, even though it has a high prevalence, is still a very serious disorder. FH is described as the cause of many cardiovascular diseases as well as of mortality at an early age. People who have this genetic disorder are expected to live 10–20 years less than the average Dutch person. The *seriousness* of FH therefore should not be underestimated:

> The total mortality caused by coronary heart diseases of men and women aged between 20 and 39 with FH is much larger than for a comparable age group in the general population. The life expectancy of FH patients is on average 10 to 20 years below that of the general population, especially if these patients are also smokers.
>
> (Kastelein and Defesche 1995, p. 7)

By emphasizing the serious consequences of FH, a distinction is implicitly made between this genetic risk and the risks for members of the general population, who, with cardiovascular diseases, still live 10–20 years longer. If many people have raised cholesterol, for most the risk of early death will not be increased to the same degree as for those with FH.

A third argument in the geneticization of FH is the very claim that it is a *genetic* disorder. 'FH is a dominant, non sex-related hereditary disorder that afflicted parents may pass on to half of their offspring' (Kastelein and Defesche 1995, p. 4). The authors of this policy plan leave no room for doubt as to the genetic character of FH. On the contrary, in the annual report of 1996, they claim: 'The molecular-genetic basis of familial hypercholesterolaemia (FH) has been established, which enables substantial improvement and simplification of the diagnostics' (StOEH 1996, p. 3). Moreover, in all reports and brochures this disorder is presented as a monogenetic disorder[7]: 'it is the most frequently occurring monogenetic hereditary metabolic disease ... The underlying cause of FH is a defect (mutation) in a piece of hereditary material (gene) that is responsible for the production of the LDL-receptor' (Kastelein and Defesche 1995, p. 7).

If FH is systematically presented as a monogenetic disease, fairly little attention is thereby paid to the fact that many mutations are involved in the emergence of FH. In 1994 it was thought that in the Netherlands, there were about 100 mutations. These accents in the presentation of FH constitute a major element in the construction of difference in relation to the general population's cholesterol problems. After all, the emphasis on the monogenetic nature of FH also produces the association of a simple and clear clinical picture. In contrast to the messy, complex, multifactorial character of the risk of cardiovascular disease among the general population, the risk of FH can be established clearly and transparently: it can be located in one gene. While we do not know exactly which interventions are most appropriate for reducing the risk in the general population – given that smoking, exercise, healthy diet, less stress and medication are each in their own way complicated or hardly effective – a monogenetic risk is associated with a *magic bullet*, an effective remedy.

A fourth way in which the difference between the straightforwardness of the genetic risk and the complexity of other risks is constructed involves the mentioning of the fact that, despite there being a myriad of mutations, these particular mutations really *do* something. It is claimed, in other words, that the mutations are almost fully penetrant and that they thus have a *large predictive value* regarding high cholesterol levels and cardiovascular diseases:

> FH is probably 100 per cent penetrant, meaning that the disorder almost always reveals itself in a raised cholesterol level ... This raised cholesterol level constitutes the most forceful risk factor for early atherosclerosis (hardening of the arteries), which in many cases leads to cardiovascular diseases.
>
> (Kastelein and Defesche 1995, p. 5)

In the first years of its existence, then, the StOEH creates a nearly linear causal link between mutation and clinical picture, and because of this link it is clear how FH should be diagnosed. If the relationship between raised cholesterol and cardiovascular disease in the general population is made to depend on the presence of other risk factors (and their weighting), in the case of FH the diagnosis is certain:

> Today in a number of cases a certain diagnosis of FH is possible by directly establishing the defect in the gene that causes the disorder.
>
> (Kastelein and Defesche 1995, p. 5)

> The establishment of the underlying defect that causes FH, namely the mutation in the gene that codes for the LDL-receptor, constitutes the actual proof for the diagnosis of FH.
>
> (Kastelein and Defesche 1995, p. 11)

In other words, while the estimation of the non-genetic risks of cardiovascular diseases is a matter of interpreting and *judging*, the genetic diagnosis of FH is presented as the *objective evidence* of the risk. For a long time, the identification of FH also counted as a syndrome diagnosis, a diagnosis based on the interpretation of multiple ubiquitous features of FH, but molecular biology has made it possible to demonstrate its single cause in a crystal-clear fashion. This implies, according to the representatives of the StOEH, that the mutation – rather than the measured cholesterol level – has to be the basis for the diagnosis. In a brochure for families we read: 'Presence of the mutation provides 100 per cent certainty that someone has FH. Conversely, absence of the mutation offers full certainty that someone does not have the disease' (StOEH 1999a). In the brochure of 2000 this formulation was dropped, but here too genetic diagnostics is called the most reliable way of establishing FH. And recently we could read on the website of the diagnostic laboratory of the StOEH in the Amsterdam Medical Centre: 'The golden standard for the diagnosis FH is finding the mutations that are responsible for the defect LDL-receptor' (www.jojogenetics.nl). Apart from the fact that genetic diagnostics is called most reliable, this method for detecting FH is also presented as 'simple'. While a cholesterol measurement has to be repeated to produce a reliable result, the DNA test is a matter of a single shot. 'This can be done through a simple blood test' (StOEH 2000). A reliable, predictable and simple diagnostic method is available, then, for this serious and often occurring disease, FH. This diagnostic technique can adequately distinguish between individuals with FH and those without it.

In the wake of this logic, the StOEH investigators make a fifth move toward geneticizing FH. After all, when the diagnosis is certain, one also has a basis for intervention. Not the actual cholesterol level, but the detected mutation has to be the *criterion for treatment*. The researchers of the StOEH argue that a mutation has already done damage to the arteries before it reveals itself in a raised cholesterol level. It is the mutation that defines FH and constitutes the occasion for starting early with treatment in the pre-symptomatic phase. Postponing treatment until a high cholesterol level is established would, in fact, amount to neglecting these patients. The mutation thus becomes the criterion for distinguishing people with a specific genetic risk for cholesterol problems and cardiovascular diseases from people with a multifactorial risk pattern, *and* for demarcating a special genetic risk group from the normal population.

A final step in this string of arguments constructing the difference between genetic and other risks involves the premise that FH is not only a clear monogenetic disorder and that there is a simple test to establish it, but that there is also an *effective remedy*. While for people with an array of risk factors for cardiovascular disease the choice of an effective intervention (giving up smoking, better diet, medication, more exercise) is a matter of 'trial and error', once diagnosed with FH it is quite clear what has to be done. Through medication and lifestyle regimens, the chance of contracting cardiovascular disease can be effectively lowered. If detected early, it is possible not only to slow down the process of atherosclerosis but also to reverse it.

This particular logic subsequently serves as the basis for articulating the problem for which the FH detection programme provides the answer. This problem is that only a limited percentage of FH patients are treated. Many – as much as 90 per cent according to the 1996 annual report of the StOEH – are still out there, ignorant of their genetic constitution, notwithstaning they could be saved with simple means. They belong to 'the highest risk category for death at a young age' (Kastelein and Defesche 1995, p. 8), but many continue to be deprived of medication.[8] The Hereditary Hypercholesterolaemia Foundation (EHC), which is geared to public information and FH research, even speaks of a large group of people with 'a "time bomb" in their body' (EHC, no date). The StOEH, the EHC and Bloedlink – the patient organization that originated in the EHC – view it as a public responsibility to protect this specific genetic risk group against the risk of premature death, as expressed quite frankly in Bloedlink's newsletter: 'still, each and every day, a handful of young people kick the bucket, while the solution is so simple' (Bloedlink 1999, p. 2). Understandably, this organization characterizes delays in the detection effort – for instance, because of a lack of funds – as 'irresponsible' and 'immoral' (Bloedlink 1998, p. 3).

The policy plans, annual reports and brochures of the StOEH allow us to infer how FH is constructed as a specific genetic disease that strikes a specific genetic risk group. In the ensuing argument, the genetic disorder FH – in comparison to the complex and multifactorial character of cardiovascular disease that affects the entire population – appears as a miracle of simplicity: a monogenetic disease, a reliable test and an effective therapy.

This definition of FH as a genetic disease does not imply, however, that those involved in the detection programme give no attention to lifestyle factors. The researchers affiliated with the StOEH explicitly acknowledge that lifestyle risk factors, notably smoking, can reinforce the symptoms of FH. In the treatment of children, for instance, the prevention of smoking is given much attention. Nonetheless, in the array of risk factors, the raised LDL-cholesterol – the effect of the defective gene – is the main one:

With the introduction of risk factors (smoking, fatty diet, obesity, lack of exercise) at the start of this century, the life expectancy of patients with FH has decreased. This is a strong argument in favour of intervention through elimination of these risk factors. The main risk factor, however, continues to be the raised LDL-cholesterol level.

(Kastelein and Defesche 1995, p. 8)

Likewise, the patient organization, in its newsletters and brochures, devotes a lot of attention to a healthy, non-Western lifestyle, fervently advocating a Mediterranean diet (olive oil, fish, vegetables, fruit and red wine) and exercise, green tea and stress reduction. The above quotation might as well be interpreted as an effort to put the genetic risk's significance into perspective. Yet this attention to the risks of an unhealthy lifestyle suggests a strengthening of the distinction between the group with the genetic risks and the general population rather than a weakening of it. For instance, in its business plan Bloedlink claims: 'People with familial hypercholesterolaemia get heart infarcts at a youthful age, even if they do not smoke, have no high blood pressure, have no diabetes and are not overweight' (Bloedlink 1998, p. 8). In short, the mutation is also active in the absence of other risk factors for cardiovascular disease. While in the general population a change of lifestyle can be effective when it comes to the prevention of cardiovascular disease, this is not sufficient in the case of FH patients. They do have to live healthily but, like diabetes patients, they cannot do without medication (Bloedlink 1999, p. 1). Following the logic of the StOEH and Bloedlink, it is hard to conclude otherwise than that detection of this group has the utmost urgency. But, oddly, not all actors appear to be convinced of this view.

Controversy

The detection of people with a familial basis for raised cholesterol levels and cardiovascular disease relies on a definition of FH as a genetic disorder caused by a mutation at the LDL-receptor gene. Thus a distinction is constructed between this specific genetic risk group and the across-population risk of cardiovascular disease. This legitimization and argumentative strategy is understandable in light of the already existing cholesterol practice of GPs. A monocausal approach is deployed against the risk-factors approach that used to dominate the prevention of cardiovascular diseases. The tension between both approaches, however, makes clear that different prevention strategies, aimed at divergent risk groups, do not go together harmoniously.

The first formal evaluation of the Dutch FH detection programme gave rise to a controversy. When allocating funds for the detection of FH for the period 1994–98, the Dutch health minister asked for an evaluation. In this context a research group from the AMC performed a cost-effectiveness analysis *and* an analysis of the social and psychological aspects of the detection of FH. The controversy centred on the former, the cost, whereby the added value of the genetic approach was explicitly raised as an issue.

This evaluation study concluded, among other things, that there was a strong, but no absolute, correlation between found mutations and clinical symptoms such as a raised cholesterol level. Seventeen percent of people with a mutation proved to have no raised cholesterol (Marang-van de Mheen, ten Asbroek et al. 2000). Furthermore, 75 per cent of those with a mutation received cholesterol lowering medication, but three-quarters of them had already been prescribed it through their GP, based on a clinical diagnosis. It was mainly young people with a mutation who, on account of the detection programme, were first put under a medical regime. The research group was not very positive about the cost-effectiveness of the detection programme, which was, it claimed, 'in the grey area' (Marang-van de Mheen Mheen, ten Asbroek et al. 2000, p. 93). Moreover, the above-mentioned estimate was too optimistic, so it reasoned, because regarding the supposed health benefit, one had assumed that all detected FH's would actually receive medication while in practice this was only 93 per cent.

In response to this evaluation report a discussion meeting was organized, where the line of argument developed by the StOEH (supported by its patient organization) – in which FH appeared as a specific genetic disorder of a specific risk group – was deconstructed. At that meeting, many themes were addressed. In the context of my analysis, the discussion about the criterion for calculating the years of life gained is the most interesting theme because the assessors thereby used the Framingham-risk function. This meant they calculated the effects of the detection, and of the cholesterol lowering in those detected, on the basis of the correlation between cholesterol and the risk of cardiovascular disease in the *entire* population. According to the StOEH and Bloedlink, this was unjust because FH would concern a specific, genetic high-risk group, where 'the Framingham function does not apply to persons with FH' (ZON 2000, p. 6). In the discussion, however, this proved to be a matter of perspective.

For instance, the thin correlation between mutation, cholesterol level and cardiovascular disease was given much more weight in the discussion than in policy reports, annual reports and brochures from the StOEH. While its representatives claimed that a mutation was found in 85 out of every 100 people with clinically diagnosed FH, which would legitimize the detection effort's continuation. This was contrasted with the following by one discussant:

> There is also a larger group with less clear symptoms. The percentage of people in this group in whom a mutation is found is much lower. Of 1500 persons who die prematurely of cardiovascular diseases, 10 to 20 have FH. A bottom-up detection of persons with a high risk of cardio-vascular disease would need to be adopted as policy; for example, a specific primary health care prevention programme (comparable to deploying a diabetes nurse).
>
> (ZON 2000, p. 10)

This participant in the discussion, then, put the meaning of the genetic risk into perspective where it concerned the providing of preventive health care: there was a larger group that could lay claim to it. The representative of theStOEH, however, held on to a genetic angle:

> The remarks by Mr Smelt are correct. But most of the 600 people who die each year of a cardiovascular disease before the age of 45 are the carrier of a mutation. It is better to begin with a top-down approach.
>
> (ZON 2000, p. 10)

In the ensuing discussion it became clear that the dispute was hardly a theoretical one. Instead, the key issue involved was of who could lay claim to care in a situation of limited means. After all, the choice for a specific diagnostic test also defines the risk group that will be centre-stage in the preventive care effort. One discussant observed that

> screening and treatment cost money and every guilder can only be spent once. It is important therefore that if society is to pay the cost, there also has to be a social benefit. Two-thirds of the population has a cholesterol of over 5 mmol/l, and in the drinking water scenario (chol-esterol lowering medication for everyone) the average gain in life expectancy is 0.6 to 0.8 years. What is the social benefit of genetic screening over cholesterol screening?
>
> (ZON 2000, p. 10)

In their reasoning the representatives of the StOEH started from the given of a genetic risk group, and thus they automatically referred to the DNA test as detection method. By contrast, many other participants in the debate started from the fact that many people die of cardiovascular diseases; to detect these risks they had no qualms about cholesterol measurement. Next, the discussion moved on to the issue of the most relevant risk group and concentrated on the most adequate diagnostic technique for identifying it. This was quite clear to the StOEH:

> A screening study of FH based on cholesterol measuring alone falls short. It turned out that in families with FH, a significant percentage (15–20 per cent) of the persons without mutation still has high cholesterol and hence they are incorrectly recorded as an FH patient.
>
> (ZON 2000, p. 11)

The discussion leader tried to remove the sting from the discussion by claiming that the use of a DNA test in the diagnosis of FH was no longer an issue of debate, but that of course was exactly what the discussion was all about. What in fact is an FH patient? Is FH the result of a DNA test and an established mutation, or are there any other reasons for diagnosing someone as having FH? And when raised cholesterol is used as the starting point, what is the difference between people with raised cholesterol and no mutation and people with a mutation and no raised cholesterol? No wonder the discussion leader's intervention was revoked immediately:

> The DNA test does not determine the diagnosis of FH. There is a group of people in whom no mutation is established. A DNA test, then, does not offer 100 per cent certainty, and a clinical diagnosis is better ... by applying the DNA test one is geared to a specific population. One should not thereby forget the larger problem, namely people with a high risk of cardiovascular disease.
>
> (ZON 2000, p. 11)

The FH patients' representative, however, argued that not all high risks were the same. There was a difference between people with a non-genetic high risk for cardiovascular disease and the FH high-risk group, and his organization wanted to hold on to the detection of 'persons with a genetic mutation [and with] a high risk of contracting cardiovascular diseases at an early age' (ZON 2000, p. 11). This was subsequently countered by the claim that measuring the cholesterol level was enough to detect them. 'But then all these persons have a high cholesterol level' (ZON 2000, p. 11).

By stressing the complexity of the interrelationship of mutation, cholesterol and cardiovascular disease, the discussion on the evaluation of the detection of FH put the significance of the DNA test in the detection of risks for cardiovascular disease in a different light. While the StOEH had given the geneticization of FH a strong push in order to support the detection of FH with the help of DNA diagnostics, in the discussion its argument was unravelled. The StOEH had underexposed the complexity of the relation between mutation and clinical picture, but in the debate precisely this complexity received much attention. One of the participants, who tried to put the added value of DNA diagnostics in detecting high risks for cardiovascular disease into perspective, suggested that it would be found in patients' larger compliance with the proposed treatment.

> There might after all be an added value in the genetic testing of these persons, because persons with such high risk are possibly more compliant with treatment, in part because of the familial character of the disorder.
>
> (ZON 2000, p. 11)

This implies, ironically, that genetics is seen as a solution not to the problem of cardiovascular disease, but to people's limited obedience in following medical professionals' advice.

If the discussion on the evaluation of the detection did lead to recommendations for further research, for quality enhancement of detection and registration, and for better coordination between GPs, the StOEH and the lipids clinics, the detection programme's effort as such was continued and even accelerated. The conceptual work in the discussion aimed at unravelling the StOEH's line of argument was insufficient to halt the detection programme.

The paradox involved is that this detection practice was made possible by the geneticization of FH, but it also provided the material – DNA, pedigrees, clinical pictures – to revise the reductionist, causal explanation of FH. As argued above, diagnostic techniques evolve through their usage. The analysis of the DNA collected by the StOEH, then, resulted in a much more complicated picture of FH than this organization had given earlier. In the 1997 annual report, for instance, we can read how mutations of another gene (LPL) can have much influence on the emergence of cardiovascular diseases:

> In 1997 the identification of additional hereditary risk factors in patients with cardiovascular diseases was started. A mutation was found in the lipoprotein lipase (LPL)-gene, the so-called N291s mutation, which further raises the cardiovascular risk in patients who have FH by a factor of 4, and therefore these people have a 16 to 20 times higher chance of dying of cardiovascular disease before the age of 60. It also

turned out that this mutation's effect was strongly influenced by circumstantial factors such as lifestyle and body weight.

(StOEH 1997, p. 12)

In a StOEH document from 1999, three genes are mentioned that play a role in FH and as many as three forms of FH are distinguished: FH1, FH2 and FH3 (StOEH, 1999, p. 10). This suggests that the notion of FH as a simple monogenetic disease was gradually adjusted in the StOEH's own research effort. Further, that the claims on the predictability of the mutations were at least premature shows from the announcement in the 1998 annual report of a large study of 500 FH patients after establishing the relationship between the specific mutation and the clinical picture, because, in fact, they may vary quite substantially (StOEH 1998, p. 16). The publication of a recent dissertation by the Amsterdam research group, entitled *The spectrum of premature atherosclerosis: From single gene to complex genetic disorder* (Trip 2002), symbolizes the developments in the research that was made possible by the detection effort. These developments trigger the question whether the effectiveness of the detection programme should not be localized more in the link between laboratory and screening than in the actual prevention of cardiovascular disease. This question, however, was not addressed in the evaluation.

Although the controversy on the added value of genetic screening failed to have specific policy consequences and the detection effort involving FH was continued, the discussion nicely illustrates the clashing perspectives on the prevention of cardiovascular diseases. It reveals that prevention strategies that are geared to pushing back the incidence of cardiovascular diseases in the general population, and that take an array of risk factors as explanation and starting point, do not mesh harmoniously with prevention strategies that target specific genetic high-risk groups and, in so doing, start from a causal relation between gene and disease. The dividing lines between the various risk groups cannot be drawn so easily, and how precisely one draws those lines depends on one's trust in specific diagnostic methods and one's specific style of thinking. Moreover, as one of the discussants put it, the money 'can only be spent once'. This is why the debate on the detection of FH does not exclusively involve scientific concerns; it also deals with the question of which groups deserve our public funds most.

Causes, risks and solidarity

In this chapter, I addressed the extent to which the introduction of molecular biology in the field of cardiovascular disease leads to geneticization.

Based on the vicissitudes of the detection of FH, it has become clear that geneticization is not the inevitable consequence of DNA technology, but that much *conceptual work* had to be done to define FH as a specific genetic disease with a clear and indisputable genetic cause. This work implied that the difference between the existing risk-factor approach and the genetic perspective was magnified. The thinking in terms of a complex interplay of a multitude of risk factors was countered by a reductionist causal explanation, while the idea that the risks of a particular lifestyle are what mainly lead to cardiovascular diseases was disqualified by pointing to the effects of the genes. While the majority of the population would bring cardiovascular diseases down on themselves through an unhealthy lifestyle, it was argued that, based on the result of DNA testing, a specific group of 'patients' can be identified who are victims of their biological fate. Even if many in this group have no health complaints, they are likely to be smitten by illness and death before long. Of course, an unhealthy lifestyle will also play a role in this risk group, but while others could drastically reduce their risk of cardiovascular disease by changing their lifestyle, so it was argued, this applies only marginally to members of this group. Healthy living cannot undo the harmful effect of mutations, but medication can. Because members of this group are not the masters of their own destiny, they deserve special care, such as, for instance, a preventive detection programme and reimbursement of the cost of cholesterol reducers.

This conceptual work also triggered a critical response from representatives of the risk-factor approach who criticized the detection of a special risk group on the basis of its genetic constitution, arguing that this group cannot be delimited from other high risks through DNA diagnostics. Their response proved to have little practical effect, however. The detection of FH was further pursued and the public began to view FH as a simple monocausal disease, even if scientists knew better.

As said, in the discussions on FH it is not only a matter of a scientific difference of opinion. The issue is which groups deserve what kind of public care. My analysis shows that this issue cannot be addressed by referring to the 'evidence' about the nature, seriousness and size of risks. That 'evidence', after all, was subject to discussion and, depending on the approach one embraces, one will present other forms of evidence. By starting from the assumption that there are genetic risk groups for cardiovascular disease which can be identified by DNA testing and that there is a causal line between genes and disease, one will argue that it is reasonable to take specific preventive measures for such groups. However, by starting from the view that cardiovascular diseases are a public health problem and emphasizing the complexity of the interaction between the various risk factors, one will propose differently organized care. Risk groups, this much has become clear, do not exist but are *made*, while the work that goes into constructing similarities and differences between risk groups (size, seriousness, screening, etc.) in order to be able to treat them differently or not is *normative work*.

Although the geneticization of FH appears to be a technical-scientific process that only seems to apply to FH, this process has implications for the *public* significance of cardiovascular diseases and for how people assign *responsibility* for health and illness. It is no coincidence, for instance, that in the discussion on the value of DNA diagnostics for identifying FH patients, the cause of FH – the mutation – is made so important. According to the social philosopher Judith Shklar, we are inclined to make solidarity dependent on the causes of suffering, rather than simply sharing the responsibility for the effects of the setback involved (Shklar 1990). Apparently, we are more inclined toward compassion if someone is smitten by a natural phenomenon than when someone gets in trouble by their own doing. Thus understood, the emphasis on the genetic cause of FH is an adequate strategy for organizing public means. But, so Shklar argues, the boundary between 'what happens to us' and 'what is done to us' is not so clear, nor is the boundary between 'what happens to us' and 'what we do to ourselves' a given. We construct such boundaries and, as such, they are political facts. If Shklar demonstrates that 'a black skin' can be considered a social phenomenon, so we consider 'genetic cause' a social phenomenon – the result of human work.

However, Shklar suggests that in our thinking about solidarity and justice, it might generally be more productive to downplay the distinction between 'what happens to us' and 'what we do to ourselves', between nature and human work. Justice and solidarity, she argues, basically imply that victims are acknowledged and helped, regardless of the cause of their harm. Therefore the focus would have to be less oriented toward the cause of the damage and more toward the damage itself. From this perspective, solidarity means that we help to repair the damage, irrespective of the question of how it was caused. Or, in even stronger terms, solidarity involves trying to make the question of the cause have a limited role in acknowledging and repairing harm. From this vantage point, the mutation carrier – no matter how seemingly clear the explanation for his risk of cardiovascular disease, no matter how simple the cause of the predicted harm – is not automatically more entitled to solidarity than a representative of a risky lifestyle – neither because the cause would be more unequivocal, nor because the cause would be more respectable.

The analysis of the controversy involving the detection of FH reveals that with the rise of genetics in the prevention of cardiovascular disease, a *new arena* emerges in which the approach of lifestyle risks to the general population and the approach of specific genetic risk groups compete with each other to win public legitimization and public funds. The lifestyle risk-factors approach offered an alternative to physiological reductionism. This approach provides much conceptual space to allow for complexity and poor control, but in recent years there has been a tendency within this approach, as argued in the introduction, to turn a risky lifestyle from a collective responsibility into an individual one (Petersen and Lupton, 1996; Van Hoyweghen 2006). Smoking, being overweight, not enough exercise – these are increasingly seen as an expression of individual failure. The emergence of genetic causal explanations for disease implies that this individualization trend still gets an *extra dimension*. If some genetic risks of people are beginning to be considered as a 'biological fate' for which they bear no responsibility, this causes other risks to be associated with the opposite: they are – unintentionally – a matter of their 'own choice', even more than before. By stressing the fact that the fate of FH patients is 'given', the fate of those without a genetic risk for cardiovascular disease (the overall population) becomes even more 'voluntarily chosen'. In this historical context, geneticization, or genetic essentialism, thus goes together with rising expectations of individual control of one's lifestyle and a larger inclination to judge people on their failing control (cf. Van Hoyweghen, Chapter 6 in this volume).

In the near future we are bound to see many discussions on the costs tied to predictive and preventive genetic techniques and the associated interventions (cf. Boenink, Chapter 3 in this volume, on the cost of monitoring in suspected hereditary breast cancer). For example, reimbursement of the cost of cholesterol lowering medication has already been discussed several times. For which cost are we to take public responsibility and which not? Which risk groups do we grant our solidarity and of which groups do we feel that they ought to have their destiny into their own hands and that thus they cannot lay claim to public money? Given the inclination to make compassion and solidarity depend on the causes of suffering, it is advisable, as Hedgecoe proposed, to enrich public debates with critical analyses of the normative work going on in processes of geneticization. Such analyses contribute to a situation in which genetic representations of causes and responsibilities will not automatically become a public legitimization for the distribution of scarce funds. It is also advisable to follow the suggestion of Shklar and attribute less of a prominent place in discussions on solidarity to the cause of suffering (or projected suffering) than to the suffering itself. Thus we will be able to allot 'genetic causes' and 'the risks of the lifestyle' the same starting position in debates on solidarity in the design of public health care.

Acknowledgements

This study was part of the research project 'Problems of social cohesion in the era of predictive medicine' which was funded by the Netherlands Organization for Scientific Research (NWO). In the context of this study, interviews were held with researchers in the field of FH, StOEH board members, representatives from the patient organization Bloedlink, researchers involved in the evaluation study, representatives from the Dutch Ministry of Health, Welfare and Sport and from the Health Care Insurance Board, GPs and internists. Although this chapter is based on written sources and these interviews are used as background information, I thank all these various individuals for their willingness to be interviewed. I also thank the StOEH and the Dutch College of General Practitioners for access to their archive and I thank Professor Dr John Kastelein for his comments on this chapter.

5
Detecting Familial Hypercholesterolaemia: Escaping the Family History?

Klasien Horstman and Carin Smand

Genetics and family bonds

Over time, family life has lost ground in Western societies. Migration, fewer children per family, the emancipation of women and children, the rise of the number of divorces and other factors have contributed to a process of individualization, which, at least in Western Europe, has caused family bonds to grow less important (NWO 2003). In response to the question 'who am I?', fewer people are likely to refer to the family in which they were raised as the defining factor, while more will point to their own particular choices and achievements. Most look at themselves primarily as an individual rather than as a family member. In combination with this development, the naturalness of the family bond and contact among relatives – out of a sense of duty or habit – has increasingly been replaced with a preference for self-chosen contacts. With some relatives, one is in touch, but other relatives one rarely meets if at all.

Despite the many examples of declining family life, it is also possible to challenge the individualization thesis. The family's reduced significance appears to apply mainly to extended family relationships. In many European countries the nuclear family is still a basic institution that largely determines with which skills and repertoires a child is equipped (Brinkgreve and Van Stolk 1997). Frequently, nuclear family relatives can rely on each other in times of serious illness or death. The sheer size of the informal, family-based care and support system suggests that at least from the angle of health care, the significance of the family as a social unit should not be underestimated (Dijkstra 2003; Potting 2001). This putting into perspective of the individualization thesis cannot obscure, however, that the meaning of family relationships has changed over the past 50 years. If the instinctive sense of responsibility and mutual concern within families has not altogether disappeared, individual choice has become the prevailing ideal in family contacts.

In light of this changed meaning of family relationships, the issue of the influence of genetic diagnostics on family relationships presents an interesting case. By definition, of course, genetic diseases are 'in the family' and genetic diagnostics forces family members to face the fact that at least they are genetically related to each other. The contribution by Boenink on genetic diagnostics of breast cancer (in Chapter 3) shows, for example, how mutually dependent members of one family are when it comes to making a diagnosis. Blood and information is needed from several family members to enable a specific family member's proper diagnosis. Conversely, a specific person's genetic testing result may have implications for the potential risks of his or her relatives. Genetic diagnostics, then, does not only pertain to individuals, but automatically implies involvement of family members.

What effect do the processes of genetic diagnostics have on family relationships? On the one hand, genetic diagnostics may well strengthen family ties. An invitation for genetic diagnostics can make individuals who have not seen each other in years realize that they are part of a single family. A shared diagnosis can enlarge the solidarity among family members and make them aware that they are all in the same boat. On the other hand, genetic diagnostics may also put pressure on family relationships. If someone is unwilling to supply a blood sample for one relative's diagnosis or if someone would rather not hear about a particular test result, the distance between family members may grow and relationships may deteriorate. That hereditary diseases are family diseases by definition does not tell us what genetic diagnostics is doing *in* and *to* families.[1]

The anthropologist Finkler has argued that in the era of genetic diagnostics the nature of family is drastically changing (Finkler 2000). Genetics confronts people with their genetic destiny and through genetic testing families become medicalized. Based on interviews with healthy and afflicted women from families in which hereditary breast cancer occurs, Finkler demonstrates that women start seeing their family through genetic glasses. Through the process of genetic diagnostics, they are reminded of family members they had forgotten and they seek contact with family members they lost sight of. The plain fact of *shared* DNA generates an awareness of family that they lacked before. Finkler concludes that medicalization has put substantial pressure on the prevailing family ideology (that contact with relatives is a matter of choice). Merely because people share the same genes, they begin to experience a sense of kinship and solidarity with others that is not necessarily based on mutual love, affection or friendship. At a time of drastic individualization, genetics thus contributes to the construction of a new family identity and history.

Against the background of the work of Finkler, in this chapter we explore what genetic diagnostics of familial hypercholesterolaemia (FH) brings about in families. As became clear in the previous chapter, since around 1990 people from families with a genetic risk of FH – and hence with an assumed raised risk of cardiovascular diseases – have been subject to a systematic detection programme in the Netherlands (Umans, Defesche et al. 2000). Via a so-called 'index patient', who is known to have FH, the Detection Hereditary Hypercholesterolaemia Foundation (StOEH) actively approaches family members with information about the fact that FH occurs in their family. They are asked whether they would like to undergo a DNA test to discover if they have the genetic disorder as well. When people opt for testing and indeed prove to carry the specific mutation linked to a raised cholesterol level, they are prescribed medication and are encouraged to change their lifestyle. The idea here is that those in whom raised cholesterol is not yet established may also have a mutation and that they will benefit from medication at an early stage. According to the StOEH, some 40,000 people in the Netherlands are thought to have a mutation that predicts FH. While the previous chapter focused on an analysis of the detection programme's *development*, in this chapter we concentrate on the *effects* of the detection – on the influence of this programme on family interactions and relationships.

Our example differs in a number of ways from that of Finkler. For one thing, most women in her study ask for genetic diagnostics themselves after suspecting hereditary breast cancer.[2] They are the ones to initiate the contact with clinical geneticists. But in the case of FH it is the nationwide detection programme that takes the initiative to inform specific people about a particular genetic risk. If for hereditary breast cancer very drastic prevention methods are available but no therapies, the motivation for a systematic FH detection programme is that one really has to offer something to people. In case of a positive DNA testing result, people may drastically reduce the risk of early mortality through medication and changing their lifestyle. Detection of genetic risks, it is claimed, makes it possible to turn high risks into normal risks.[3] In other words, detection of FH provides people with the chance to escape their genetic fate and break from their family history. The success of the detection programme, then, does not lie in the specific value of the DNA diagnostics as such, but in its actual use: a positive testing result opens up the road to a normalization of the risk. The detection programme thus reflects the notion 'knowledge is power': power to face up to diseases and early death. While Finkler shows that genetic diagnostics leads to enhanced risk awareness and to the medicalization of family relationships, the FH detection programme bolsters the notion that it is possible to turn the tide of one's family history: it allows those with FH to escape their biological fate. In this chapter we examine whether this is in fact true in actual practice.

With the collaboration of the StOEH, the Dutch foundation that organizes the nationwide FH detection programme, we have conducted 19 semi-structured interviews with members from five families that were offered genetic diagnostics and also made use of it. We call them, respectively, MacDonald (four members), Peters (three members), Jones (two members), Anderson (seven members), and Taylor (three members).[4] We talked to more than our 19 interviewees, however, because sometimes others joined in our conversations: a partner, a sister, or, occasionally, a child. Two interviewees from, respectively, the Peters family and the Taylor family, rejected the offer of genetic diagnostics. We would have liked to talk to more family members who did not take up of the offer, but two family members who rejected the offer of having a DNA test also refused to participate in our study. They told us that they neither wanted to get involved in genetic diagnostics, nor talk to us about it. A third person was unwilling to participate in our study because he felt that genetic diagnostics was basically a private affair. Our sample, then, does not have an equal distribution of proponents and opponents of this detection programme. The family members we interviewed generally evaluated the FH detection programme positively.

Our investigation concentrates on the families' *experiences* with FH detection. What does detection of FH bring about in families? To what extent do genetic diagnostics allow somebody to escape their family's biological fate? It will become evident that the promise of genetic testing plays a role, in particular, in the decision of parents to have their children genetically tested. It will also become clear, however, that this promise attributes too much power to genetic testing. Inasmuch as children who tested positively can ignore the family history, this is the result of the *work* of parents and children rather than the *effect* of the test. Finally we will address to what extent the detection of FH gives rise to the development of a special solidarity among family members. But, first things first, we start with the role of family members in the detection process involving families.

Detection is family work

A nationwide detection programme for a genetic disorder implies that individuals can be confronted with information on a hereditary disease in their family of which they were entirely unaware. Because some may experience this as threatening and because the right of not-knowing in the context of genetic risks carries a lot of weight, the advantages of such detection programmes have to outweigh such concerns. This is why the detection programme capitalizes on its crucial role in successful prevention. Those who have themselves genetically tested and prove to have a mutation that is tied to FH may substantially reduce the risk of cardiovascular disease that is associated with a high cholesterol level by means of medication and change of lifestyle. A person who has a mutation that is detected at a young age and treated well has a normal life expectancy. As StOEH founders and researchers, Kastelein and Defesche, write on their website:

> For someone with hereditary raised cholesterol who is optimally geared to treatment, the risk of heart infarct is the same as for someone who is 'healthy'. In other words: a well-treated patient has a normal life expectancy and does not have to refrain from any activity or occupation.
>
> (www.spreekuurthuis.nl)

Thus the detection programme's legitimacy is couched in terms of its therapeutic consequences and the normalization of the genetic risk involved.

Strikingly, in the families we interviewed there was almost no sense of surprise when faced with information of a hereditary disease. The letter or phone call from the StOEH with the offer of a genetic test rarely came out of the blue because most of our interviewees had already been informed by other relatives, or had heard indirectly that such testing was going on in the family and that there would be a call or letter at some point:

> It was a family contact. A younger cousin visited and he told me that I would get a letter, and indeed, two days later the letter came in.
>
> (Taylor, B)

> I knew something was going on. My uncle C. had told me that someone might call, so I wasn't surprised at all.
>
> (Anderson, G)

Family work is of great importance for organizing the detection.[5] Although the offer of genetic diagnostics formally comes from the StOEH, the detection programme's success in part depends on family contacts. Preliminary work done by family members may lessen the anxiety caused by the first letter or stimulate relatives to accept the offer of genetic diagnostics.

In one family, for instance, one person tried to track down all family members, uncles, aunts and cousins to inform them in a letter about hereditary hypercholesterolaemia and the FH detection programme:

> I had already written to my relatives because it turned out that my son did not just have hypercholesterolaemia. His other problem was an extremely high LPA level, which is in fact worse, but at this point there is no remedy for it. I had that result and thought that perhaps the whole family suffered from it ... This is why I wrote a letter to the whole family to inform them and that perhaps they should have it tested ... When my grandma was still alive, this issue was hardly discussable. I used to meet my various cousins at her place, but after she died I never saw them anymore. If I continued to see some uncles and aunts, it was at my dad's birthday – he passed away in 1994 and then you lose contact ... I gathered all their addresses and also tracked down my grandpa's brothers and sisters to find yet more addresses, because I felt that everyone had to know that there was LPA ... Several aunts called me shortly thereafter and told me they were very enthusiastic about my effort. I have written my uncles and aunts asking them to inform their children. In response I heard only positive things.
>
> (Anderson, B)

These active intermediaries are of great importance to the detection programme. In some cases, warnings may be initially ignored or one will not convince family members right away of the relevance of genetic diagnostics:

> I said that my cholesterol was too high and that we may all have such a high level ... But they were headstrong. What you hear is: 'Why? I feel fine ...' They did not want to listen ... Then suddenly a male cousin died, at age 36 or something ... and then the ball started rolling ... My female cousins became afraid and set things in motion ... Next, one after the other opted for testing after all.
>
> (Taylor, A)

Although for years the work of this intermediary was hardly productive, it contributed to the fact that the family had a frame of reference which later on made it easier to identify a cousin's sudden death as a case of FH, after which the other family members soon opted for genetic diagnostics.

Some family members, however, do not like others to take on the role of detection programme partners and, for instance, pass on names and addresses to the organization:

> The only one who was indignant about it was one of my sisters. That was also the sister whom we suspect not to have it ... She told me on the phone that she was not happy that I had given her name to the StOEH. She was really troubled to have to face the fact that possibly she suffers from a disease.
>
> (Anderson, B)

The detection programme for FH will sometimes confront unsuspecting people with information on hereditary disease. With the projected beefing-up of the detection effort,[6] chances increase that in the future, individuals who do not have a relevant family history will be informed about FH in the context of cardiovascular disease. There is a chance that these people will become more angry or frightened and, in part as a result, reject the offer of DNA diagnostics. Because the FH detection programme not only defines its success in terms of the number of offered DNA tests, but also in terms of detected risks, it has little to gain from such reactions. The StOEH is therefore highly dependent on family work, the sustained effort of family members who take on the responsibility to supply the StOEH with information about as yet undetected family members and to inform and potentially convince their relatives to participate. These family members are indispensable intermediaries between the detection programme and families.

But not everyone, it turned out, welcomes this mediating effort. Although the detection programme and family members can be considered as *partners* in the programme's implementation, potential angriness and indignation about mediation efforts are primarily geared to family members. What to the StOEH is a possible anonymous case of FH who rejects the offer of DNA diagnostics, to intervening family members is an angry cousin or a foolish sister to whom they have to relate. If family members decide to play a role in a genetic detection programme, then, not just the health of their relatives is at stake but also the quality of family relations. Precisely at a time when family bonds are a matter of choice, unsolicited information or advice – no matter how well intended – can be experienced as unacceptable meddling. That this hardly proved to be an issue among our interviewees has to do with the prior history of the families involved.

It runs in the family

In the families we interviewed, in many cases the offer of genetic diagnostics proved to be no surprise because family work had already prepared them. Nor was information on a potentially too high cholesterol level news to these families. Many interviewees already had a long history of cholesterol problems and/or experience with cardiovascular diseases. Medical complaints, disease and early death already constituted major family history ingredients:

> I did not know I was a carrier of the gene, but I did know I had too high cholesterol.
>
> (Peters, A)

> I was about 13 when it was found out. My mother got a heart infarct, and then the children had to be examined, and my sister and me as well.
>
> (MacDonald, D)

To these detected people, then, the result of the DNA test did not involve a change of therapeutic regime; they already used cholesterol reducers and continue to do so.[7]

In four of the five families, there was a more or less articulated awareness of a familial defect in relation to cardiovascular diseases prior to coming into contact with the offer of DNA diagnostics. These families were not medicalized by the information on genetic risks and the offer of genetic diagnostics – they had been medicalized already. These family members already defined their family in terms of the diseases they were smitten with:

> It had long been known that heart disorders run in the family. We knew it. You do not yet know if it involves high cholesterol, but you know it is hereditary ... You already more or less know that there is no escaping.
>
> (MacDonald, A)

> Many of his children have it. I also have sisters who don't have it. You simply say it runs in the family. Around the corner there's a relative who had the same surgery as I had. We are the MacDonalds family and that explains it.
>
> (MacDonald, B)

> I saw our family doctor for a severe cold ... Then I proved to have a cholesterol level of 12. Our doctor asked if it occurs more often in our family.

I said, yes, I believe so, heart and arteries, it comes from my mother. He told me that my brothers and sisters better go in for a test. And then they all did. Out of five, three had it, of which mine was highest.

(Anderson, C)

It varies in our family; one person has it, another one does not. I knew that probably my father had it; he died of a heart infarct. We have always suspected that he had a too high cholesterol level, but it was just a guess. My mother did not have it, but I did suspect that there had to be some hereditary basis.

(Peters, A)

It is striking that the interviewees who did not make use of the offer to have a genetic test were also still thinking about it in terms of a familial defect and heredity. For them genetic diagnostics was in fact superfluous, as they knew quite well what was going on:

At that point it was known already that in our family there is too high cholesterol. If now we first hear about a particular gene, we already knew that in the Peters family high cholesterol occurs regularly. It was a familiar phenomenon, and in this respect nothing has changed.

(Peters, B)

This person regularly had his cholesterol checked by his physician, but he had ethical-political reasons for not doing a DNA test.

I felt it to be an infringement on a private situation. Who do they think they are ... this is going too far. I see entire families branded, as it were. You may think that you are living healthy, but you're wrong. You will get a disease. You have the gene. That, I feel, is beyond the pale ... Privacy is very important to me. They should not impose too much from above. You also see in newspapers that yet another study of this or that has been done, which makes you wonder if they're not pushing the envelope. Is not the entire Dutch population ill or unhealthy by now?

(Peters, B)

The other person who refused DNA diagnostics also testified to the experience of a family disease and used medication to keep her cholesterol down. Given her frequent contact with the health care system, she found it too much of an emotional burden to undergo genetic testing:

> I thought I was done for a while with the medical scene; I have had it for now ... I knew that this was a family issue; that my grandma died of a heart attack and an aunt ... At that time – in the 1960s, I was still a child – my mother was called in. Her whole family had to be examined ... And at one point, some seven years ago, I had my cholesterol measured; it was 33 per cent too high ... When I had it checked later on again, it was fine, so I thought I would leave it at that for now.
>
> (Taylor, B)

Both refused the offer of genetic diagnostics, not because they did not want to know that a raised cholesterol level and cardiovascular diseases occur more frequently than usual in the family, but precisely because they have known it for a long time. They have taken precautionary measures and genetic diagnostics does not change anything as to how they define the disease or their therapeutic regime.

The awareness of a familial defect that causes cardiovascular diseases that many of the interviewees had before crossing the path of genetic testing was expressed, among other things, in the way in which they referred to themselves in phrases like, 'I belong to the *Andersons*', 'typically one of *MacDonalds*', 'we are after all the *Taylors*'. They characterized the history of disease and mortality in the family as a family trait. 'And then we again all met in the crematorium, and my sister reacted by saying what a miserable family we have, another one gone again, so young ... Yes, we have a miserable family' (Anderson, E). In these families, then, a positive result of the genetic test scarcely proved to change the family image.

> Because we already had our share of heart afflictions, you more or less know that you won't get away with it. But now it has a name.
>
> (MacDonald, A)

> No, I always have had a metabolic disorder, as I call it; something is not produced in my body. Well, that will be the missing gene; it is called differently now, but it has the same effect in how I experience it, so I am not really surprised.
>
> (MacDonald, D)

> Don't ask for the result; yes, I already knew that it is hereditary.
>
> (Taylor, C)

Only one family in our sample first became aware of the fact that 'it runs in the family' through the offer of DNA diagnostics. Up till then the cholesterol problems of several family members were considered as more or less random facts rather than as symptoms of a hereditary disease. Only through the detection programme and the ensuing DNA testing did they began to redefine several symptoms in various family members as symptoms of a shared hereditary affliction.

In four of the five families, then, the idea of a hereditary defect that causes cardiovascular diseases was already part of the family identity. This is why, for most interviewees, the test hardly implied an intensification of their relations with the medical circuit: they already used medication and regularly had their cholesterol checked; they were patients already. Although one might expect the FH detection programme to confront people with shocking new information of a hereditary disease, in the case of our interviewees the programme hardly changed anything. They had long suspected a familial defect causing cardiovascular diseases and corroboration of their suspicions did not make them panic.

The fact that the offer and usefulness of genetic diagnostics seemed of little significance to most family members in our sample basically implies that, so far, the success of the detection effort may have been limited. Their raised risk had already been established, mostly by their GPs. For our interviewees, then, the detection programme provided no means to escape their family history. They viewed themselves as part of a family with a hereditary defect and had reconciled themselves to this fate. Inasmuch as they could do something about it, they had done so already. To them the detection programme was, as some explicitly put it, superfluous.

Familial disease, individual responsibility

The FH detection effort implies that in more or less the same time frame various family members have to decide on whether they want to accept the offer of having a DNA test, while those who decide to do the test also receive its result more or less simultaneously. Research by Finkler has shown that such a shared trajectory in the case of hereditary breast cancer occasions new relationships within the family (Finkler 2000). Did this also happen in the families we interviewed?

It turns out that most family members framed the issue of whether or not those in the same generation – brothers, sisters, male cousins, female cousins – should have themselves tested in terms of choice and individual responsibility. Meddling is rejected because by and large they believe it to be a matter of personal decision. In practice, however, this issue of autonomy is addressed differently within families.

In the Peters family, for instance, the individual responsibility of all siblings explicitly counted as a major value. But this did not imply that there was mutual indifference. On the contrary, in this family difference of opinion on the meaning of DNA testing went hand in hand with regular contact and care for each other. Most of the siblings, for instance, failed to grasp why one brother rejected DNA diagnostics:

> We're all living quite near to each other and have quite intensive contact and we have talked about it. To me it was astonishing, for instance, to learn that my brother, whom you talked to, said he was not going to go in for a test. We call that playing the ostrich.
>
> (Peters, A)

This brother, however, considered it a matter of his own responsibility:

> I do not know whether all my brothers know it, for we are a large family (ten children). But most know that I will not have myself tested; they know it ... It is a matter of seeing each other regularly and occasionally you talk about the issue. But I consider it to be a matter of my own responsibility.
>
> (Peters, B)

While others believed his rejection of genetic diagnostics not to be sensible, they agreed with the idea of it being a matter of his own responsibility:

> My brother worries that in this country they are heading for a situation where at birth you get a genetic diagram that says you will only reach the age of 40. I find it extreme ... We talk about it, but at one point it is clear that he has his point of view and I have mine and that's it then; you cannot change his stance and he cannot change ours ... You can indicate why, give examples of how relatives have fared, but you can do no more ... I have listened to his view, but, for me, I just want to know how things stand.
>
> (Peters, C)

In this family the idea of individual responsibility not only applies to decisions on genetic diagnostics, but also, and mainly so, to the meaning of health in general. Even if it involved one sister's risky lifestyle, family members tried to avoid their being concerned from being perceived as meddling:

> My oldest sister, we feel, has bad lifestyle habits, and we also talk about it. All my sisters have very good contact with each other and

each year they go out together and then they automatically talk about this kind of topic ... it is mostly in a pleasant atmosphere. I will not lecture her ... I am also concerned that she has too high cholesterol and is overweight ... of course I worry about it, especially because we have such good contact ... As it happens, next week I see her and then I will talk to her about my having this interview ... I know how much she struggles with the issue herself.

(Peters, A)

This is very important in our family, individual responsibility ... Telling a brother to mind his drinking or whatever is as far we go, but I wouldn't interfere more. He has to decide for himself. It is also because you are part of a big family. It happened that at one point one of my brothers needed support after all, and then you try to help out of course. But otherwise it is up to him.

(Peters, B)

As such the Peters family is perhaps a prototypical modern family. Although there is mutual solidarity, the vocabulary of individual choice and responsibility is highly valued. Differences of opinion are not left unaddressed, nor are they avoided, but the relatives do not pressure each other. In this family, discussion does not result in coercion, but in accepting mutual differences of opinion. They try to avoid these differences from becoming centre-stage in familial interactions. If individual freedom is not more important than the family bond, leaving each other free is important for maintaining good family ties. Jogging together, which all the brothers do, matters more than insisting on each other's different view on genetic diagnostics.

Yet differences of opinion on the role of DNA diagnostics may also go together with limited social contact between family members, as well as little mutual involvement. This is the case in the Jones family. In this family, the differences in dealing with FH were not the effect of genetic diagnostics, but an expression of already existing differences in attitude to life. Those who shared a more or less similar outlook also talked about FH diagnostics. Where this shared attitude was absent, however, FH diagnostics was left unaddressed: accepting their mutual differences and going about their business. If there are no large conflicts, there is relative detachment vis-à-vis family members who have a different lifestyle.

I do not live by the day, but I do feel that I will not grow very old, so I go on vacation, I become a little more relaxed, do not save all my pennies for later ... My brother is somewhat like that. He rides his bicycle to work, is very driven and in good shape, watches his diet and goes to bed early. It makes me feel he has no joy in life. I once said that he should see a cardiologist rather than an internist. Then he says, yes, you always know better. Well, that settles it; with one brother or sister your contact is different than with another.

(Jones, A)

With my sister-in-law, whom you talked to, I can talk about it very well and she is also very open about it. She has also had her daughters tested ... But contacts with my husband's younger brother are all but non-existent; in fact, he never contacted us and this is still true, nor did he ever want to have much to do with it anyway. You notice that he is unwilling to participate. When I meet them at a birthday party I'd better not raise the issue.

(Jones, B)

The Anderson family, finally, illustrates that little interaction on FH is expressed not only in accepting mutual lifestyle differences, but also in frictions and irritations. If there was or is much disagreement on many issues, discussion of FH diagnostics dances to the same tune.

There is no discussion. Most to whom I wrote a letter I have not seen. If you meet each other at all, it's at funerals ... There are old sores here and there. It has to do with illnesses you also had to cope with ... A few things have gone wrong at funerals in the past, which is why some do not interact anymore ... When I look around I see many families that have some problem. You notice that emotional events trigger blockades and cause relationships to disintegrate.

(Anderson, B)

One of the other family members and his wife confirm this picture.

Earlier in fact the family interactions came down to zero as well. As a family we are not very close ... There are other people with whom we can talk about it, like friends and acquaintances.

(Anderson, D)

These three families demonstrate three different patterns in dealing with the offer of genetic diagnostics. These differences have to do mainly with the specific history of those families, the established forms of conduct and the mutual expectations among family members. One family tries to create room for both individual freedom and mutual involvement: differences of opinion on genetic diagnostics are explicitly discussed, but each family member is allowed to have a view of their own. In another family, individual responsibility means that family members with various lifestyles have gone in different directions and as a result they have little to do with each other anymore. Individual experiences with FH are primarily discussed with family members who share the same lifestyle. In a family with a history of mutual fighting and irritations, it turns out that the issue of FH diagnostics gives rise to frustration and reproaches as well. As such these stories show that families, once genetic diagnostics becomes an issue, do not start with a clean slate. On the contrary, the events concerning DNA diagnostics merely underline the nature of the family relationships. This is why the offer of such diagnostics hardly affects the existing family patterns.

Aside from such established patterns and their differences, in all the families studied we noticed evidence of processes of individualization. That family members have expectations about mutual care and attention, and that they are disappointed when they are not met, shows that family bonds still hold special meaning. But the families in our sample also demonstrate that mutual contacts exist by virtue of an affinity in attitude or outlook on life. Family members may prove to share a defect gene, but if they were not in touch with each other to begin with, this genetic knowledge does little to change the situation.

The promise of testing

Most of our interviewees learned hardly any new information upon receiving their genetic test result, nor did it change the nature of their family relationships or their sense of their own overall health. Still, they proved to be quite interested in knowing exactly what is going on. Precisely in the light of their family's history of cardiovascular diseases and early mortality, most interviewees expressed the view that it is worth knowing whether or not FH is involved. 'You'd better know it. They also say it is better to start with medication early' (MacDonalds, A).

It turned out that they seemed more concerned about their children's destiny than their own. While they spoke about their hereditary defect as a more or less given, viewing themselves as an integral part of the family history, they welcomed genetic diagnostics for their children as a means to turn the tide of that particular history. Genetic diagnostics is seen as a way to gain control over FH in the youngest generation and allow its members to escape the family's biological fate.[8] Consequently, most interviewees had the DNA of their under-age children tested as well.

There has been much debate on the application of genetic tests in children.[9] On the one hand, it raises the issue of children's privacy: they have to be able to determine on their own what they want to know about their genetic constitution and what they do not. On the other hand, genetic diagnostics is regarded as an opportunity to optimize the care for the child in case of a positive test result (Clarke 1998). Against the backdrop of these debates, some physicians have argued for restraint regarding FH detection in minors. Given the uncertainties in the interpretation of positive results, the age of 18 would be early enough for detection, thus allowing children to have a *normal* youth (Gevers Leuven 1997). The StOEH, however, relies on research that shows that a mutation is not just potentially harmful in the future, but that it causes damage to the arteries from the very start. Even if there is no raised cholesterol level, the arteries still gradually become clogged up (Bakker, Wiegman et al. 1997; Wiegman, Rodenburg et al. 2003). The detection programme's message, then, is: 'The sooner one knows, the better.' In children who test positively lifestyle can be taken into account right away and if needed one may also start with medication.[10] Our analysis, however, is not geared to deciding a normative position on the question whether it is right that children are tested for FH. We are primarily interested in the views and experiences of the interviewees regarding the effects of this testing.

Given the detection programme's message ('the sooner, the better') and our interviewees' own experiences with cardiovascular disease, it is hardly surprising that none of them voiced the idea that it would be better for their children to remain ignorant about the potential mutation associated with FH. Similarly, the norm of individual responsibility that applies to family members from the same generation and that might be a reason for postponing diagnostics in children to a later age played no role in parents' considerations about whether their children should be genetically tested. Nearly all interviewees emphasized the importance of 'knowing', so as to be able to take adequate prevention measures in case of a positive result.

For instance, a woman whose husband has died, and whose three children aged between 15 and 20 have been tested, commented as follows:

> I hardly made any considerations, in fact, I simply said to myself 'I will do it, I want to learn more about it'. Because, as you can imagine, it is something the boys might have as well. It allows you perhaps to do something about it at a young age.
>
> (MacDonald, C)

Likewise, others were positive about testing children:

> The two of us always felt that it should be done. When my wife raised the issue with our family doctor, he said that it was not necessary yet, that they are still too young and that when they get to the age of 10, 11, 12, we should just come by his office. So we simply waited until they were that age.
>
> (MacDonald, D)

> It makes it possible to prepare them quite early and adapt their lifestyle to it. Children do not take it into account; they just go out to get a large fries. But when you know you have something hereditary, you can take it into account early on.
>
> (Peters, A)

> I feel that if we bring children into this world who are predisposed by a defect gene, they have to know this ... I wouldn't be able to imagine what the burden might be in a case of FH.
>
> (Anderson, F)

Even the woman who wanted no DNA test thereby referred to the fact that she had no children. Had she had them, she would have felt it her duty to have a test herself:

> That I have no children also played a role. I discussed it with my sister. She said she has a daughter who also has two children, and that she wanted to know if perhaps she had passed it on. It makes me feel, well, that I don't have that responsibility.
>
> (Taylor, B)

The other person in our sample who refrained from testing did not raise the issue of genetic detection screening with his children. He is the only interviewee who felt that parents, when it comes to decisions on DNA testing, should not interfere in their children's autonomy:

> I discussed it with my wife, but not with our children. They know that every now and then they need to have themselves checked. In this area the same applies: it should not be imposed, not by organizations and certainly not by parents.
>
> (Peters, B)

These children, however, were grown-up already and regular screening for cholesterol was seen as a normal thing to do in this particular family.

It is made fairly easy for parents to decide to have their young children tested as well. The StOEH's general information aimed at children relies on a more or less standard explanation of FH as a 'shortage of fishing rods that fish for the cholesterol in the blood'. The organization of its detection effort is such that children can more or less automatically join in the parents' testing trajectory because the StOEH staff take blood samples from participants at their homes. It is only a small step, then, to tell children what is happening and also take a blood sample from them:

> Well, all three of them were there when my blood sample was taken and then they start asking questions ... Next, we simply talked to our children and asked them what they wanted, a sample or no sample. The two eldest said right away that it was no problem, but the youngest proved to be rather fearful of the needle.
>
> (Peters, A)

> I discussed it with them, albeit very concisely, and waited for the StOEH visit, thinking that they can better explain it. So they were sitting here around the table, and they explained once more exactly what it was all about. And then, yes, you can still decide whether or not you have yourself tested ... The story was actually very clear and there was also the curiosity of 'do I have it or not'.
>
> (MacDonald, C)

In light, on the one hand, of the experiences with illness and death in the family, and the promise of DNA testing on the other, most parents we interviewed felt it to be their duty to consider their child's risks when offered a test. Because cholesterol problems 'run in the family', genetic diagnostics was fairly normal to the parents themselves and in the relative security and intimacy of the home situation, the actual taking of a blood sample becomes a quite unobtrusive issue. By having their children tested, the parents believed they were taking their responsibility by providing a favourable turn to the family history and saving their child.

The work of parents and children

Our interviews revealed that parental hopes regarding genetic diagnostics in children run high because the test may result in a break from the family history. In the practice of everyday life, however, both parents and children have to put in *work* to make that expectation reach fruition. This has already begun with the decision in favour of testing.

In 'the sooner, the better' logic, the practice of DNA testing for FH in children becomes a straightforward and non-controversial element of a curative trajectory, rather than a predictive technique that after careful consideration one may decide to apply.[11] As such there is little or no negotiation on the genetic testing of children, as is true in other forms of diagnostics when children have medical complaints. Parents are the ones to take the decisions in this respect. When their children are slightly older, there may be more room for negotiation, but it still does not take place from the view that it is up to the child to decide. Some children simply go along with a specific decision. One father told us about his adolescent son: 'It was my wish to know it ... I urged my son to have a DNA test, so as to spare him the troubles I had to go through' (Anderson, B). The son confirms this picture: 'My father told me about it. He knew a lot about it ... My parents told me to have myself tested and so I did' (Anderson, A). The mother of another family with small children explains:

> After Saint Nicolas we explained that dad has a clogged artery and that when fathers have this, their children may also have it and that is why they need to be tested. This means that a blood sample would be taken. If you have it as well, it means that we have to eat as healthy as possible.
>
> (Anderson, D)

Sometimes parents and children have different opinions on testing. The parents, given their medical family history, are convinced of the usefulness of genetic tests, but the children, who do not share that history in the same way, may have another point of view. In that situation, they need to be persuaded to participate in the testing trajectory:

> It was simply like we'd better know ... I did ask myself whether I should get Pete involved in the whole shebang ... I made him decide for himself, inasmuch as this is possible at 12. But I felt I had to do it ... Charlie did not want his DNA tested; it was extremely hard on him. I talked him round.
>
> (Jones, B)

In the families we interviewed we encountered no case of parents who refrained from genetic diagnostics because their child resisted it. Occasionally, of course, children protested, but in those cases also parents were eventually the ones to decide. Some proved to be quite inventive. For instance, the mother of Pete and Charlie got the idea of honouring both her own wish to know whether her son had FH and her son's wish not to know the result by not opening the result letter together.

> I said that I would like to know it and if he doesn't that I will be the one to open the letter so that I know it and can tell him whenever he is ready for it. I also said that it is important if later on he wants to raise a family of his own. When we were discussing it he asked me to tell it to him after all … No, he never again said anything about why I talked him into it.
>
> (Jones, B)

In some families, the two parents proved to hold different views on genetic diagnostics for their children. For instance, the parent with a family history of cholesterol and cardiovascular disease urged testing more emphatically than the parent without such history.

> To you it was all obvious and clear, but for me as partner it was a bothersome period. I thought suppose our two have it, then I am the only who doesn't. I managed to handle it, but it was really difficult. If I was realistic enough to think that if there is medication, we can benefit from it, I also felt like: if we don't know it, it does not hurt us.
>
> (Anderson, D)

The parent without such a family history clearly had more trouble with genetic diagnostics than the parent with the history. These examples suggest that in families in which there are no huge differences of opinion on the usefulness of DNA testing in children, this testing may still be experienced differently. These various perspectives and experiences show that decisions on genetic diagnostics for children may cause frictions in the relationship between parents or between parents and children, and that work is needed to make sure these frictions remain manageable and local.

When the results come in, there is also work to be done. Most parents considered their own DNA test result hardly a burden; rather it was a confirmation of what they knew already. Given the promise of prevention, a positive result for a child should not have to be a source of great concern for parents. They then know, after all, what must be done to make sure their child stays healthy. A positive result can even be seen as marking the beginning of a new family identity – one that is no longer characterized by disease and death from cardiovascular disease. Yet parents tended to take the genetic testing results of their children more seriously than their own results. They anxiously waited for the news about their children's fate to arrive in the mail, the thickness of the letter symbolizing the result:

> The anxiety went up only after we got three letters in the mail at the same time. There was one thick letter and two that were not as thick, so I thought two are positive or one is ... It turned out to be a relief for the two, but the third, who opened the letter later on, said: 'What am I to do with it?'
>
> (MacDonald, C)

A negative result made those involved feel relieved of course. 'We were extremely happy that none of our three kids has it. What a relief' (Peters, A). And as one child commented: 'Relieved! ... I'd have felt more burdened if I had it and would have to take medication. In fact I would have hated it' (Peters, A junior).

When hearing that their children tested positively, most parents were quite shaken. One mother said that for a long time she could not view her child otherwise than as a 'walking heart infarct'. It is even harder on parents when their various children have different results:

> It was terrible to learn that one of the three had it, because it makes you think and things resurface again, how it all went, and why he died so suddenly; yes, it is hard on me.
>
> (MacDonald, C)

> And then the results came in and it turned out our youngest had it ... It made me feel a little sad ... at one point the oldest asked whether the results were in already. 'Yes, we got them. You don't have it, but your brother does.' Then he said to feel happy about it. To what extent they understood it is serious I don't know. He felt like he didn't have to cut back on mayonnaise ... Then our youngest came in who said: 'Oh, no, I have it as well.' We were sitting around the table, the three of us, and he was the one to have it and the two of us didn't. ... I would wake up at night thinking that he is the one to have it ... it sounds objective perhaps, but there is also the emotion ... He does have to swallow a capsule each day. Of course it is possible to defeat the disease, but I would rather have a healthy child.
>
> (Anderson, D)

The husband: 'Yes, such a result is really a blow' (Anderson, D). For most children, too, it was hard learning that they tested positively. Three children were so bothered with it that they were unwilling to talk about it with us. 'I asked all three of them; they don't like it, so I leave it at that' (Jones, B).

Parents and children do not simply consider a positive result of the DNA test a fact, based on which they can start preventive therapy. Although in psychological studies of children with FH it is concluded that these children's quality of life does not go down significantly,[12] a positive result is a burden to both parents and children. Rather than the beginning of recovery or staying healthy, it is an indication of unhealthiness. If a child with a positive result who receives proper treatment is statistically seen as a normalized risk, from the angle of the family history parents experience their child as a raised risk. A positive result reminds them of the illness and premature death of other family members and these memories are projected onto the child. For the time being, and despite all the therapeutic options, a child who tests positively still fits into the larger family history.

In many cases a positive result will turn a child into a patient, rather than into a 'normal risk'. This is seen in particular regarding worries about that child's lifestyle. Sometimes the parents involved felt that their child did not sufficiently take into account the positive result and then developed a lifestyle that, according to the parents, was unhealthy. 'Our middle one wanted to start smoking. She suffers from high cholesterol, so we put in a lot of effort to keep her from doing it' (Jones, A). In another family within the same extended family, one child proved to be very concerned about his positive result and even became obsessed with a healthy lifestyle and started behaving like a patient:

> We discovered that Pete also has it, and he uses medication now. It is true that we pay attention to it, but with young children – he is only 12 – you should not go too far. He developed a dislike for eating … Since his father died and since he also knows that he has it, he asked with every cookie he got: 'should I really eat it?' … He actually doesn't want to go to his grandparents anymore because they give him a lot of candy and other sweets … So it has a large impact on him.
>
> (Jones, B)

In these cases parents worry and put in much effort to regain a proper balance for their child. Even if there were hardly any immediate problems, parents still registered their child's reaction to his positive testing and they had to deal with it. For instance, one father told of the changes in his 16-year-old son, who tested positively:

> I suspect that he tries to seize the day even more. Being out on the town comes first for him, but well, the same is true of his friends. He has become more of a bon vivant ... I have the idea that when he is at home he sits in his room more often, cutting himself off from others a little more.

(Anderson, B)

A mother from another family explained how she struggles with the fact that her children hardly eat vegetables and with the contradictory advice for her son with epilepsy who also proved to have FH: while eggs would be good for those with epilepsy, if you have FH it is better to eat fewer eggs. While one parent decided to talk often about FH and the importance of a healthy lifestyle, another mother decided not to talk too much about it. She suspected that this kind of conversations would result in resistance to the pressure felt by children to adopt a healthy lifestyle and would precisely lead to unhealthy behaviour.

The FH detection programme is the expression of fairly straightforward prevention optimism: the sooner the better. This optimism, which mainly emphasizes the benefits of detection and not its toll, motivates parents to have their children's DNA tested as well. In practice it turns out that the process of genetic diagnostics puts more pressure on the mutual relations within the nuclear family than the promise of 'normalizing the risk' appears to suggest. There is resistance with children, sadness about positive results, and concern about how children respond. After a positive result, one does not move on to the order of the day indifferently. On the contrary, a positive result is a nasty result. While such a result means a success to the programme's detection effort, the family has yet another potential patient. If the programme's job is done, the parents' work has only just started.

This should not have to lead to the conclusion that the promise of a genetic test as the road to normalizing risks is not realized. It may teach us, however, that this promise attributes too much power and responsibility to the genetic test while obscuring from view the trajectory *following* the test result. After all, it is not the DNA test that turns children with a raised risk into a normal risk; this work has to be done by the parents and children involved. If parents opted to gain specific information about their child's DNA, a positive result should not come as a total surprise. Yet its implications are rather hard to anticipate: parents after all cannot predict how their child or they themselves will respond to that result, for only time will tell. A *clear* positive result, then, marks the start of an *uncertain* trajectory. The very *knowledge* that parents wanted to have also confronts them with the *responsibility* to make sure that within that trajectory the child with a positive result (and with lifestyle advice and, possibly, medication) does not eventually become a patient as well. A positive testing result implies an additional challenge to parents and children. They are the ones who subsequently have to put in the effort towards normalizing the risk of FH.[13]

Familial disease and the boundaries of the private sphere

As our argument suggests, the idea of individual responsibility plays a major role in dealing with FH within families. But when the testing of children is involved, things are a little more complicated. Decisions on their testing are basically taken within the nuclear family and are primarily the responsibility of parents. But parents who support DNA testing in children and have their children tested still find it difficult to accept that other family members reject genetic diagnostics for their own children. They feel that these parents, so our interviews suggest, deny their children's right to be informed. The ease with which family members leave each other free to decide on DNA testing that exclusively concerns themselves sharply contrasts with the difficulty many have when parental decisions on testing have particular consequences for their children, especially of course when these decisions do not agree with their own views or decisions.

In the reaction of families to such differences of opinion we may again discern three patterns. The individual in the Peters family who did not want a genetic test for political-ideological reasons felt that parents should not take this kind of decision for their children. 'It should not be imposed, not by organizations and certainly not by parents' (Peters, B). In this family this view was explicitly discussed.

> It is something we talked about with the other brothers and sisters in the manner of 'if you do not have it tested yourself, your children might not know it either and information is withheld from them.'
>
> (Peters, A)

Concerning the weighing of the right to information on a genetic disorder in the family versus the right of children to decide on their own about genetic diagnostics, however, the Peters brothers and sisters failed to reach agreement.

If most people do not want to interfere with decisions from other parents within the extended family about their children, their health is still a concern to them. This interviewee's story shows that she was unable to swallow the fact that her nieces and nephews were not tested:

> If you know you have it, there is a large chance that one of your children has it as well, and you may help this child ... Her mother told me, for my sister-in-law did not tell me directly, but my mother-in-law regretted it that her daughter did not have her children tested ... We have a family day soon at which the nieces and nephews all meet each other; it is not quite the occasion for raising such things ... It is a pity that those children have not been tested, as is also the view of my mother-in-law ... but my sister-in-law apparently has good reasons for it ... You can do something about it ever earlier, so I think that what you actually can do you also should do.
>
> (MacDonald, C)

If this interviewee did not yet dare to raise the issue with her sister-in-law of having her children tested, nonetheless she kept cautiously looking for a good opportunity:

> It is of course hard to raise it, for you do not know how exactly she feels; you have to wait for the right moment. And it may take some time before the occasion presents itself. ... I suspect she wants to defend herself, but for me she doesn't need to. I am just curious whether she has decided to have their children tested after all, or whether she's fed up with it; I do have my suspicions ... Actually, I find it abnormal not to do it. In my opinion if something is offered to you, to have yourself tested, either you have something or you don't, that you always have to do it ... If it happened to a close relative, you take that offer to have your children ... I mean, you don't want to lose someone ... I don't want her to think like, 'that is none of your business; it is my family; these are my children.' Look, this is why I am very cautious; it is basically up to her, but I hope that she will think twice and one day will say, 'I will do it after all.'
>
> (MacDonald, C)

In the Jones family, one did not go to such great lengths. Although parental decisions not to have children tested met with lack of understanding, family members put in no effort to discuss it because they did not expect it to be productive. In this family, the ideal of individual responsibility mainly seemed to justify that one should not worry about it any longer:

> If she did not have herself tested perhaps I would have told her to do so in relation to her children and grandchildren ... to bring it under her attention for once. But I would not have kept going on about it, for it is her responsibility.
>
> (Jones, B)

> He does not want to participate, nor have his children tested; this is not smart because he has extremely high cholesterol, he knows it, but well, we never really talk about it, he always has to leave early, this is how he adapted his life. I once told him, hey, why are you doing it this way, but he does not really reply ... What a pity in fact, for probably one or two of his children have it as well.
>
> (Jones, A)

In the Anderson family, finally, the quarrelsome atmosphere also comprised the issue of their children's health. The parents were more bothered by indifferent reactions from family members when it involved genetic diagnostics in children than when it pertained to them:

> *Mr*: We also had a lot of quarrelling about it [testing of children] and no one asked if the results were known already ...
> *Mrs*: I feel that it is too hard a topic to discuss ... it is the nature of the family. In his family it is tough; they are willing to talk but only about things that go well, vacations and the like ... I told one of his sisters that it was difficult for me. And she understood.
> *Mr*: But she never came back to it.
> *Mrs*: No, you are right. Your other sister did phone to tell that she was relieved that hers didn't have it. But she never inquired after those of us.
>
> (Anderson, D)

Earlier irritations and differences of opinion take on an extra dimension when one feels rebuffed where it concerns the health of one's children.

In the above we have seen that family members consider decisions on genetic diagnostics and one's lifestyle a matter of individual responsibility. They leave each other free. Although they do not give up individual responsibility as a basic standard when decisions on genetic diagnostics in children are at stake, it takes more effort to act on it. The worries about the children of family members prove to be much larger than the worries about those family members themselves: one finds it hard to bear that a niece or nephew may have a disease and that the parents are doing nothing about it. The ideal of individual responsibility, however, makes it difficult for parents who are positive about DNA testing in children to express their view vis-à-vis parents who decide differently. Are they allowed to say something about it or not, and if so, how?[14] The notion of a hereditary disease in the family does not automatically make it more legitimate to interfere with each other's children. In fact, the cherished ideal of autonomy rather implies that much work is needed to articulate one's worries and uncertainties about someone else's children in such a manner that existing family relations do not suffer. FH in the family results neither in a putting into perspective of the ideal of individual responsibility, nor in intensified family relations or removal of the boundaries of the private sphere. Raising and caring for children first and foremost remains a matter of individual parental couples and the remaining space for criticism and solidarity proves to be limited.

The history of the family is stronger than the family history

The FH detection programme holds the promise of early detection of FH, adequate therapy and the normalization of the risk. In this sense FH detection may imply a break with the family history. This promise proves to play a role, in particular, regarding the testing of children. While the older generation family members have long known they belong to a family with a hereditary defect that causes cardiovascular diseases, which is why DNA diagnostics has little added value to them, they view genetic diagnostics applied to their children as a way of breaking with the family history. They embrace the offer of DNA testing because their children, in case of a positive result, may still have a normal life expectancy. The test is seen as a road to normalization of the risk.

In the practice of everyday family life, however, things appear to be a little more complex. A positive result in a child does not by definition reflect a breaking point in the history. The test has all but done away with uncertainties. If all children test negatively, the trajectory can be completed with relief, but with one or more children with a positive result, a new trajectory begins. Parents and children, so it turns out, have to put in a lot of work to avoid the child with a positive result becoming a patient, an outsider in the family, or a 'walking heart infarct'. Talking or not talking about it, snacks on the table or no snacks on the table, formulating anti-smoking contracts et cetera – depending on their knowledge of their child, parents have to decide their norms on raising and caring for children. Whether parents – and children – succeed in dissociating children from the family history, only time will tell. Whether a detected mutation becomes a decisive biological fate depends in particular on the social work that has to be put in to allow carriers to live a *normal* everyday life. One may submit that 'living healthy' is important for each child and that the parental challenge is part of their normal task of raising children. But our interviews suggest that fixing and maintaining rules for candy and snacks for children with FH is still experienced differently than for children without health problems. It takes work to avoid FH from becoming a burden and to make sure that in a balanced way FH is part of the educating, caring for, and interacting with children – it takes work, in other words, to address FH and not make its presence too much felt. Parents and children will have to learn this along the way.

In this effort to *normalize* FH, however, parents and children prove to be left to their own devices within the larger family configuration. Finkler argued that DNA diagnostics, in the case of breast cancer, makes people face up to their biological destiny and that this causes the emergence of new family relations and new forms of solidarity within the family. Whether this is a product of the fact that Finkler attributes too much power to genetics in changing family patterns, we cannot assess. In the detection of FH, at least, something else appears to be going on. In the families we studied, we saw that in their mutual interactions the ideal of individual responsibility was centre-stage and that one was mindful of respecting the privacy boundaries involved. The offer and application of genetic diagnostics does not change this. While the new genetic tests for FH imply much new work for parents and children, they lead neither to new family patterns nor to new forms of solidarity. In the case of FH, in other words, the history of the family proves to be stronger than the family history.

Acknowledgements

This chapter was part of the research project 'Problems of social cohesion in the era of predictive medicine', that was financed by the Netherlands Organization for Scientific Research. We thank the StOEH and all the interviewees for their willingness to participate in this study.

6
Genetics and Insurance: New Technologies, New Policies, New Responsibilities

Ine Van Hoyweghen

Exploring new technologies and risks in medical underwriting

One of the most discussed topics in public debates on genetics is the use of genetic testing in insurance. This is hardly surprising, of course, as the world of insurance concerns us all. Yet the growth of predictive medicine, and genetics in particular, raises a number of new questions. Will genetic knowledge make some people uninsurable and will this lead to the emergence of a 'genetic underclass'? What about the privacy issues involved? Should asking for genetic information be prohibited by law? Insurers have suggested that such prohibition would be disastrous for their business because they need to be able to have access to the same information as their clients. Insuring so-called 'burning houses', after all, will soon make them go bankrupt, they say.

The public debate on these issues has been rather provisional and abstract (Ewald 1999; Van Hoyweghen 2004, Van Hoyweghen, Horstman and Schepers 2006). As a result, discussions failed to move beyond rhetorical slogans – 'genetic underclass', 'burning houses' – and frequently ended in a deadlock. In an effort to find new and more specific openings, I have studied the insurance world *from the inside*. My research concentrated on the work life insurers have to do in order to underwrite insurance applicants. How do insurers assess risks? Which considerations and techniques are thereby used? And what are the consequences for insurability? In this chapter, the focus is specifically on how insurers deal with predictive medicine, notably the role of genetics and lifestyle, in medical underwriting.

My argument relies on fieldwork I did in the period 2001–03 in the medical underwriting departments of two Belgian life insurance companies.[1]

The underwriting department is in charge of the risk assessment of insurance applicants. Based on medical information about the applicant, the insurance company establishes a premium that reflects his/her mortality risk. The insured are divided into several risk groups: standard premium, raised premium (substandard), or excluded from insurance. This assessment is done by medical advisers and lay underwriters. The medical information is obtained in various ways. Basically, all applicants are asked to fill out a medical questionnaire. Depending on the applicant's profile, they will be asked for additional information. A medical expert may perform another examination of the applicant, laboratory tests are done, and/or reports from attending physicians or specialists are requested. My study centred on mapping the 'risk trajectory',[2] the process from policy application to either acceptance (against a specific premium) or exclusion. I also opted to study interests and relevant connections within the whole company, all having one or another link with underwriting, such as the claims, actuarial and marketing departments and corporate management. An analysis was also made of the various instruments used in the underwriting process, such as computer programs, statistics and reinsurance handbooks. Finally, based on interviews and written sources, I describe the overall context of life insurance by considering the particular role of specific Belgian regulations, the professional associations of insurers, the medical advisers, the actuaries and the patient organizations.

Before addressing the underwriting departments and their dealings with predictive medicine it is useful to qualify the meaning of predictive medicine and genetics in Belgium. This country was one of the first European countries to introduce a legal prohibition on the use of genetic data by insurance companies. The 1992 Law on Insurance Contracts (LVO) regulates this issue in its Articles 5 and 95. Article 95 introduces a total prohibition on the use of genetic tests for predicting the future health status of policy applicants. Article 5, Section 1 of that same law requires insurance clients to supply accurate data on all information that is known to them and which they feel could be in the interest of the insurer to know. It is not permitted, however, to pass on genetic information. Belgian law also prohibits the use of genetic tests with the aim of potentially benefiting applicants. They are, in other words, not allowed to pass on 'favourable' genetic information in order to obtain a lower premium. Although this issue is also formally regulated in the Netherlands, in the 1998 Medical Examination Act (Wet Medische Keuringen, WMK), the Dutch government mainly opted for regulation by insurers themselves via a moratorium.[3] It indicates that basically, insurers in the Netherlands are not allowed to request hereditary testing as a condition for taking out insurance, but above a certain limit (the so-called 'ceiling') the policy applicant does have to provide information derived from hereditary testing.[4] In contrast to Belgium, then, in the Netherlands the ban on hereditary testing in insurance is not total.

Legal commentary on the Belgian legislation was, however, critical of this, and pointed to a lacuna in the law's interpretation. For one thing, Belgian legislators did not define 'genetic data' (Freriks 1994, p. 28), or 'genetic research techniques' (Nys 1992, p. 216) – as is true of similar legislation in other EU countries. Consequently, it is not clear whether the legislator currently forbids the use of family histories or genetic information derived from routine medical examinations, like blood tests, for example. Furthermore, its sloppy design and formulation, some legal experts argue, has caused a lack of clarity as to whether this law does or does not apply to *all* predictive tests for determining a person's 'future state of health'. In this sense, Article 95 distinguishes between 'current state of health' and 'future state of health', the latter applying to genetic testing techniques. Nevertheless, insurers have always used other (non-genetic) prognostic procedures and risk factors (such as smoking) that can also be seen as predictive of someone's future state of health.[5] There is ample room, then, for confusion.

Another concern is that, in a way, insurers always engage in a predictive effort. If genetics in public debates is often presented as 'new' or 'trail-blazing', we should ask ourselves what, really, is so new about these developments. After all, medical underwriting in life insurance has always been geared to assessing the (future) mortality risk of policy applicants. In other words, the business of life insurance centres on statistically assessing an individual's mortality risk. The basis for this risk assessment is defined as the policy applicant's 'pre-existing condition'. It is true, though, that the current developments involving predictive medicine make it theoretically possible to detect ever-more risk factors. This raises the issue of how insurers deal with 'predictive medicine'. Are predictive risk factors seen as a matter of already 'pre-existent conditions' and, as such, taken into account as independent, decisive factors for assessing mortality risks? Or are they constructed as dependent predictive traits that may or may not contribute to a person's future diseases? The introduction of 'predictive medicine' in insurance is therefore better understood as both continuous, and yet discontinuous, with traditional underwriting. Given this lack of clarity, it has been a priority during my fieldwork to observe how predictive medicine is mobilized, defined and constructed by the involved actors. How do insurers deal with risk factors that have to do with lifestyle and genetics?

Lifestyle as predictive health information

In the course of my fieldwork it became clear that underwriters pay a lot of attention to applicants' lifestyle traits or their lifestyle-predictive health information. In checking the information on policy applicants, for instance, they indicate that they mainly encounter 'lifestyle risks' or 'diseases of civilization':

> Recently we have noticed a lot of depression. And also diseases of luxury such as increased liver values, high blood pressure, increased blood sugar and so on. All this has to do with ... stress and poor lifestyle habits. These are all things, it seems to me, that can be avoided. But they have become the most common, which is too bad in a way.
>
> (Case 1: underwriter E)

The instruments and forms used to request this information also mirror this attention to lifestyle. The medical questionnaire has a separate rubric devoted to questions on weight, blood pressure, alcohol use, smoking behaviour and drug use. Its heading, 'major information', has a grey frame to emphasize the rubric's significance. During my observations it became clear that the underwriters attach a great deal of importance to the responses in this rubric:

> These questions [points to questionnaire and reads] 'weight, height, smoking behaviour and alcohol use', we consider carefully because to us these are major risk selection criteria. This is why we always return the questionnaire when an applicant has failed to respond to these questions.
>
> (Case 1: underwriter P3)

Observation of an underwriter who goes through the questionnaire and reads:

> Someone of 46, a man, assistant film director, nervous breakdown, late last year to beginning this year, at home for seven to eight weeks ... Well, a depression/nervous breakdown, it automatically means that we ask for more information. Or take stress; its mere mention is enough for us to pursue it.
>
> (Case 1: underwriter K)

The studies and statistics that underlie the premium calculation also reflect this attention to lifestyle. A recent actuarial study, for instance, listed the following 'diseases' that are associated with excessive mortality: being under/overweight, alcoholism, hypertension, smoking, depression and heart conditions. Following this study, the department's management decided to institute a higher premium for these features, but at first it hesitated to do so exclusively on the basis of this kind of health behaviour. After all, if this lifestyle information is obtained via the questionnaire, it could easily lead to faulty data and fraud on the part of the applicant. To address this concern, it was eventually decided to calculate higher premiums only if the same information was provided by more 'objective' instruments, such as blood analyses, liver tests, lung X-rays or codeine tests. Thanks to these technologies, lifestyle features can be 'measured' and thus serve as a legitimate basis for assigning higher premiums.

Similarly, the reinsurance guidelines on cholesterol, obesity and high blood pressure emphasize the relevance of predictive lifestyle information. Where these risk factors represent a statistically increased chance that someone will develop a particular disease, they are considered *primary* mortality risks in insurance, as independent bases for assigning a higher premium. Thus they are assessed in the same way as other 'pre-existing conditions' like tuberculosis or asthma, as may be illustrated through the example of raised cholesterol levels. A high cholesterol level comes with an increased risk of clogged arteries or other cardiovascular diseases. In insurance practice, however, high cholesterol is synonymous with an increased mortality risk. This same logic can also be seen in how obesity and high blood pressure are considered. Apparently there is a tendency in the insurance industry to reify predictive lifestyle factors into autonomous mortality risks. These lifestyle risks are regarded as 'prior-existing conditions', or as prior-existing 'damage' or 'deviation'. In the case of high blood pressure, for instance, an underwriter explains:

> High blood pressure is not so bad as such. You do not die from it immediately. But the heart of someone who does not pay attention to it suffers a lot, becomes larger, the muscles weaken. This is how it is with many things. There is always a chance that problems will occur later on. So to us, this is already an increased mortality risk. We have to look at the long-term effects.
>
> (Case 2: underwriter R)

Insurers thus transform epidemiological risks (defined as probabilities for future disease) into diseases or deviations that already exist. Probabilistic risk factors thus become physical abnormalities or deviations that require exclusion or a higher premium. Moreover, through the construction of these lifestyle risk factors as 'deviations', the perspective shifts to intervention and prevention. In this way, the individual's responsibility for his or her own health is emphasized. People with lifestyles detrimental to their health are thus seen as 'already ill', and even as responsible for their illnesses, which is why they have to pay higher premiums.

Aside from the use of lifestyle factors as primary mortality risks, these elements also play a role in the classification of those who are already suffering from an (other) disorder. For instance, lifestyle may play a role in *adjusting* the statistical average of higher premiums for a specific disease. By requesting additional information, via a report from the attending physician or an examination by a medical expert, elements come to light on the specific *circumstances* of the disease. These are taken into account as prognostic factors (+/−) in assessing individual rates. Thus it is possible to trace personalized, clinical information, including, for instance, the beginning of an illness, periods of relapse, the course of the illness, response to treatment, and the causes. This is yet another way that lifestyle elements can be calculated by underwriters. Consider the following underwriter's comment regarding someone who indicated on the questionnaire that she was suffering from diabetes:

> So she fills in 'diabetes' on her questionnaire, but that is a little vague. I don't know how bad it is and so on. So I send an extra questionnaire to her physician to find out her last blood readings and to see if she has stabilized. We also want to know whether she controls her illness with insulin or just with pills and is it effective. All this, then, makes a difference. Thus we may potentially lower or raise the diabetes premium for this client.
>
> (Case 2: underwriter K)

In the case of high cholesterol, underwriters may request readings of tests performed at various intervals in order to assess whether the person involved has regularly used his or her medication for stabilizing the cholesterol level:

> Here we have a letter from the GP with tests that cover the last two years, and as you can see [points to the rubric in the letter]: the blood values are fine throughout. So, I suspect this man takes his pills regularly, because I do not notice any extremes in the values. In other words, he is controlling his illness well. This is why I will accept him against a better premium.
>
> (Case 1, underwriter B)

In other words, we are dealing here with 'compliant behaviour' on the part of the client. In this respect, the subject of the assessment is not only the medical risk involved, but also the '*moral* risk', the patient's reliability and his or her way of dealing with the disease. The applicants' premiums are fixed on the basis of their sense of responsibility for their own health. If they are *good* patients, they are rewarded with a lower premium, but if they are *disobedient*, their premium will be higher.

The policy on smoking is particularly indicative of the prominent role of lifestyle in medical underwriting for life insurance. As has been illustrated elsewhere (Van Hoyweghen, Horstman and Schepers 2006, ch. V), smoking was used in the medical underwriting process as a risk classification factor for charging smokers a substandard premium. In this regard, the reinsurers' statistical studies all point to smokers as a major category. A recent European study indicates that in the past smoking was undervalued in premium levels, both as a risk factor as such and in combination with other disorders. The study subsequently concluded:

> Since smoking has a crucial impact on the mortality of both normal and medically substandard risks, all life insurance proposal forms should ask about smoking habits, and the resulting data should be recorded for statistical purposes and should be adequately taken into account in the rating guidelines.
>
> (Swiss Re 2002, p. 10)

Moreover, smoking has of late become a factor in the calculating of the standard premium as well. If the standard premium used to be put together on the basis of non-medical elements, such as age, insured capital and sex, smoking has been added as a factor:

> Before, there used to be a standard premium for smokers and non-smokers combined. And the smokers had to pay a higher premium. But in 1999, the managers said: 'Well, non-smokers clearly have a lower risk. We will reward them with a lower standard premium.'
>
> (Case 1: underwriter K)

In practice this means that at the start of the application process, the standard premium is calculated on the basis of smoking. If it turns out that applicants are non-smokers, they will get a reduced standard premium. But if they smoke, the higher 'standard' premium must be paid. And if they are 'heavy smokers' (defined in the guidelines as: 'more than 2 packages of cigarettes per day'), they have to pay an additional premium. During the risk assessment process, then, normative ideas on individual responsibility play a major role. Basing standard premiums in part on 'smoking' behaviour rewards the non-smokers. When asked, the managers explained that this was a way of pointing out to clients that they are responsible for their own health. Aside from being a strategy for penalizing unhealthy smokers, this has also become a strategy for attracting healthy clients:

> Instead of just having unhealthy people pay extra, we chose the strategy of lowering the standard premium in the case of non-smoking. Thus we explicitly suggest to our clients that their lifestyles matter. If they do not smoke they are now rewarded via a reduced standard premium. It is, of course, a positive strategy to first assign a client the smoker's rate, and when it turns out he doesn't smoke, you can tell him that he qualifies for a reduced premium. In cases of the reverse, when someone claims to be a non-smoker but the cotinine test establishes that he is in fact, a smoker, it requires us to inform him that he has to pay a higher premium. The first strategy is more customer-friendly.
>
> (Case 1: manager J)

Essentially, the new policy reinforces the difference between a healthy and an unhealthy lifestyle.

Making the normal deviant

The introduction of lifestyle predictive information in insurance practice raises questions about the significance of the standard premium in insurance. After all, the reliance on predictive risk factors (and innovative technologies) contributes to the idea of 'the worried well', the idea that we are neither really ill nor entirely healthy (Harris 1994, quoted in Davies 1998, p. 149). In insurance practice, however, we see that those who display risky behaviour are already seen as 'ill'. Ever-more conditions are being tied to standard insurance rates. Predictive information thus causes more and more health characteristics to be applied during the underwriting process, which comes with the risk that the margins for being defined as 'in normal health' are becoming increasingly smaller. What insurers define as 'standard health' is ever-more limited. Accordingly, the normality concept in insurance is shifting from a condition of *absence of illness* to a condition of *risk resistance*. In a way, predictive knowledge immunizes the normal standard premium against disease by expanding the indicators for mortality. Ironically, although over recent decades our average life expectancy has steadily increased, the norm for being accepted as 'standard' for an insurance policy has risen over that same period of time. Thus the *norm* in insurance increasingly deviates from the *average* health status. This raises the question of whether the introduction of predictive lifestyle factors has caused the standard premium in insurance to reflect rather a 'more-than-standard-norm'. Or does it imply that our social norm is shifting toward a stricter definition of health? On the one hand, these tendencies can be seen as a continuation of the common insurance logic. After all, insurers have always engaged in predicting people's health status. On the other hand, the increasing usage of predictive knowledge and techniques evidently creates new precedents for insurance practice.

'Behave': assessing the capacity for self-control

As has already become clear, potential clients are not only examined medically, but also increasingly on their moral conduct and their attitude toward their own health. The emphasis on lifestyle in insurance has caused insurers to begin stressing applicants' ability to control their own risk. Health is thus linked to the notion of 'good citizenship'.[6] By 'measuring' lifestyle variables, one can trace to what degree individuals exercise control over their own health. If they cannot demonstrate this, they are financially 'punished' with higher premiums. Individual control over one's health is thus translated into a selection criterion for insurance policy eligibility. Poor 'body upkeepers' have to pay higher premiums. The moral claims that are thus linked to specific applicants allow insurers to contribute to the construction of the *voluntary* character of 'lifestyle' risk liability. The identification of fault and guilt serves as the basis for penalization. Where disease is, however, not a matter of fault but of fate, an ill person is seen as a *victim*. But insurers have increasingly begun to note the role of individual responsibility and they figure that those who do not take responsibility simply have to pay:

> We increasingly see people who are overweight, with high blood pressure, and diabetes. These are the main disorders today. And the heavy smokers of course ... This is counterbalanced by the cancer cases, for instance. They in fact are the real victims.
>
> (Case 1: underwriter K)

The construction of the self-inflicted nature of lifestyle risks also returns in the way underwriters deal with remissions or modifications. It is permitted in specific cases to adapt a premium after a given timeframe. Insurers, however, only do this when it concerns 'real diseases', rather than lifestyle diseases, such as obesity:

> In the case of poor lifestyle habits we will not issue an adjustment. If one used to have poor lifestyle habits this can no longer be entirely erased. There is always a specific lifestyle that automatically affects one's future. So this is not an issue. There are only a few serious diseases that return after a certain period. Of these you can say, okay, as in the case of breast cancer; when it appears the disease has been stable for ten years, we may issue a premium adjustment. So then we redetermine the premium because with breast cancer, it is not a consequence of lifestyle, it is beyond one's will. In these cases the disease is to blame. Meanwhile, lifestyle habits are dependent on people's decisions, and so in those cases we do not change our earlier decisions. Because if they used to have poor lifestyle habits, there is no guarantee that they have permanently changed their behaviour.
>
> (Case 1: medical advisor E)

> Overweight people often say: 'Yes, I weigh too much, but from now on I'm going to do something about it.' So we get a lot of reactions like these, and questions like: 'How much am I allowed to weigh so I can get the average premium?' This is not how it works of course. You can decide to lose weight, but there is no guarantee that their weight will stabilize.
>
> (Case 2: underwriter O)

The same happens when underwriters are faced with the postponement of a decision. In the case of a pregnancy, for instance, the medical examination is postponed until after birth. For those with high cholesterol, however, postponement is impossible. These patients instantly receive an increase in their premiums:

> In the case of high cholesterol we will not postpone our assessment. We go right to a higher rate. With pregnancies we could do the same, but, well, I feel it involves an issue whereby those who are pregnant are being punished for something they have no control over.
>
> (Case 1: underwriter R)

As these examples illustrate, the introduction of predictive health information goes hand in hand with a distinction of the moral significance attributed to risks. According to Petersen and Lupton, this leads to a widening of the risk concept: a *moral* risk scale comes into being, 'a continuum of moral judgement' (1996, p. 115) that extends from risks that arise beyond our control to risks that are purely due to negligence regarding our health. The insurance practice thus consolidates cultural judgements that are tied to specific risks. Such thinking reverses the cause of personal tragedy, which makes it seem that the risk has always been present as a risk, which might have been avoided if one had put in an effort to do so. As Lupton writes: 'The experience of a heart attack, a positive HIV result ... are evidence that the ill person has failed to comply with directives to reduce health risks and therefore is to blame for his or her predicament.' (1993, p. 430). And: 'Failing to protect oneself from this kind of "internally imposed" risk is understood as an individual moral issue, highlighting personal failures or weaknesses'(Lupton 1995, pp. 89–90). The calculation of extra premiums for people with poor lifestyles, then, constitutes the basis for an *own-fault* approach. Those who can prove 'good conduct' regarding their health will be rewarded for it, but those who cannot will have to pay higher premiums.

The involuntary character of genetic risks

How do insurers deal with predictive *genetic* information? As indicated above, this is very much determined by Belgian legislation that prohibits the usage of genetic information. As a consequence, underwriters only rarely encounter genetic test results or genetic information (on, e.g., Down's syndrome, cystic fibrosis, Huntington's disease). Underwriters in these cases solicit the advice of the reinsurance company. Since Belgian law does *not explicitly* prohibit the use of a family history, the two insurance companies where I did my fieldwork asked for family history data in their medical questionnaire, while the underwriters could further deduce information on serious genetic and familial disorders from reports presented by the attending physician or specialist.

However, my respondents proved to be reticent in taking this information into account in risk assessments. For one thing, they did not consider it necessary to send back medical questionnaires where the responses about family history had or had not been incompletely filled in. Moreover, this family information was only meant to confirm or negate data on diseases from which the applicant himself was already suffering. As in the case of a person with cardiovascular complaints:

> If we establish symptoms for him that have to do with heart problems, we also consider the family history. It is possible, then, [grabs guidelines]

where heart infarcts have at least occurred in the family, that we have to charge a higher rate. But if a person indicates that his father suffered from a heart affliction, while this person has no such problems at this point, we do not charge a higher premium.

(Case 1: underwriter E)

The applicant's current pathology determined whether or not family history should play a role. A *tainted* family history might raise the average and already higher premium for a specific disorder (60 per cent instead of 50 per cent). For instance, as one underwriter explained with respect to familial hypercholesterolaemia (FH):

When the applicant indicates a family history of FH but he himself has no raised LDL cholesterol levels at this point, he will be accepted against a normal premium. But if he proves to have high cholesterol levels and if we can also prove that there are specific relevant conditions in the family, these things are added up and he will end up paying a premium that is slightly higher than the average raised premium for FH. And if he discloses no family history and there are also high cholesterol levels, he will end up with the average higher premium for cholesterol.

(Class 1: underwriter P)

Although family history is taken into account, then, this element plays less of a role than lifestyle. In general, the insurers make no distinction between applicants with a good or bad family history per se; these applicants are not penalized just because of their family histories.[7] Thus, in contrast to lifestyle, predictive familial risks are not reified to autonomous, primary mortality risks. A family history of disease does not automatically make a person a high risk.

These examples demonstrate that insurers attach less significance to family history than to lifestyle factors. Underwriters also articulated normative objections against the use of family history as a decisive risk factor.

We cannot afford to give a higher premium merely on the basis of family history. We cannot tell our clients that they'll have to pay more because of their father's heart problems. This is not client-friendly. If someone honestly declares that he is not suffering from anything and our tests confirm that there is nothing wrong with his heart, charging him a higher premium is hard to defend.

· (Case 1: underwriter E)

Again we see that moral connotations are embedded in the underwriting process. Charging somebody more because of his genes or his family's medical history alone is considered insensitive or unjustifiable, because having 'bad genes' is something this person cannot help. The same objection is advanced more fiercely when it involves behavioural diseases among family members. As an underwriter noted in the case of alcoholism:

> If we consider whether the mother and father have an alcohol problem? This is an even more delicate matter. You have to realize that when we charge someone a higher rate because his mother was an alcoholic, that leaves a bad impression commercially. After all, can this person be blamed for his mother's alcoholism? So why penalize him for his mother?
>
> (Case 1: underwriter P)

Family history or 'bad genes' are not considered to be a matter of choice or self-control. In other words, this is how the *involuntary* character of genetics is constructed.

Risk carriers versus risk takers

If we compare the underwriters' approaches to predictive risk factors, lifestyle risks are far more pressing with regard to risk assessment than genetic risks. In the first case, the risk pool will not subsidize the applicant; he must bear his own risk. In the latter case, insurance companies are often willing to take the risk. Consequently, insurers construct the voluntary or involuntary character of, respectively, lifestyle and family history risk factors.

According to Petersen and Lupton (1996), the moral judgements involved in predictive medicine create on the one hand 'at risk people', that is, people with risks which are perceived as completely out of individual control, and, on the other hand, 'risky selves', or people whose risk derives from their ignorance or lack of self-control. The same tendency seems to occur in the Belgian insurance industry: lifestyle *risk takers* are treated differently than genetic risk *carriers*. This results in a fault-based approach to underwriting: Risks are assessed differently according to whether they are a result of the applicant's own fault or not. In the first case, you are made a culprit, in the latter, you are considered a victim. Implicitly then, insurers disseminate moral judgements on the responsibility for one's own health. Although both lifestyle and family history are predictors of an individual's future health status, lifestyle has gained ascendancy in the risk calculation process.

This further clarifies how insurance is a normative technology and practice. Normative judgements are disseminated in insurance on who 'deserves' solidarity and which criteria citizens have to fulfil to be included as members of the insurance pool. The introduction of predictive medicine in underwriting contributes in this regard to new standards for the distribution of responsibilities between the applicant, the insured and the insurer. On a wider front, such practices reflect how we, as a society, consider the criteria for solidarity with the sick.

As indicated in Chapter 4, the 'discovery' of genetic knowledge is associated with 'genetic essentialism' or 'geneticization' in our thinking about disease (Kitcher 1996; Lippman 1993). The chapter's author claims that genes are often attributed an overly deterministic role. This means there is a tendency to distinguish people along genetic characteristic lines and to categorize diseases into genetic and non-genetic ones, resulting in different levels of responsibility attributed to genetic and non-genetic diseases. So where genes are linked to fatalism or lack of control, lifestyle is associated with individual control or responsibility. This is illustrated by a recent comment in the *British Medical Journal* on the fault-based approach that such geneticization of disease entails. The author asked whether the discovery of genetic defects in particular individuals does not automatically make those people *powerless*. He argues that a geneticized approach brings these people 'to a learned and licensed helplessness' (Smith 2002). The new biomedical distinctions implied by the 'discovery' of genetics thus result in normative effects in the distribution of responsibilities. The same approach is found in Belgian insurance companies. The Belgian legal prohibition on using genetic information in insurance policies can be considered the institutionalization of 'genetic essentialism'. Where genetic risks are seen as 'fate', as 'involuntary' or 'uncontrollable', lifestyle risks are considered as self-induced, voluntary and one's 'own responsibility'. The outcome is a financial solidarity or collective responsibility for the genetic *risk carriers* – the collective risk pool is prepared to pay for them – and individual financial responsibility for lifestyle *risk takers* – they have to pay for their risk-taking themselves via higher premiums.

The question arises, however, to what extent such normative distinctions in assigning responsibility for disease are desirable or tenable and how these criteria can be determined. Which characteristics can we identify as controllable or uncontrollable? (Shklar 1990) What does it mean to have control over your health? In other words, the geneticization of disease and the associated fault-based approach present us with major challenges. In the next section, I address this in more detail by discussing the effects of the Belgian ban on the use of genetic information in the insurance industry.

Side-effects of a genetic essentialism

As we have seen, Articles 5 and 95 of the Belgian Insurance Law (LVO) are formulated to prevent genetic discrimination in insurance. In this respect, the Belgian law can be viewed as the crystallization of the above-described 'genetic essentialism'. By thus treating genetics separately, it is emphasized that genetic information should be characterized as fundamentally *different* from non-genetic information. This particular distinction, however, creates some major side-effects. The legal imposition of a 'wall' between genetic and non-genetic information prompts new questions, notably about the workings of general insurance policy as well as our society at large.

One of these side-effects has to do with the emergence of new forms of discrimination. For instance, Rouvroy notes:

> a discrimination between, on the one hand, people who are carriers of specific 'predispositions' or 'genetic susceptibilities' and, on the other hand, those who are carriers of mere pre-symptomatic signs of specific disorders, which are not identified via genetic analysis but via routine clinical tests.
>
> (Rouvroy 2000, p. 601; my translation)

In the same vein, Fontaine specifically criticizes Article 5 by suggesting that it 'also introduces a distinction to the disadvantage of the persons who have a disorder of a non-genetic nature that is hard to justify' (Fontaine 1999, p. 297).[8] This can be elucidated by numerous examples I encountered during my fieldwork. Take, for instance, George, who while filling in his insurance application is in perfect health but is also carrier of a genetic mutation for colon cancer, which, according to Belgian law, he was not allowed to mention. Next, there was Peter, whose health was as good as that of George's but who via routine tests was diagnosed for specific pre-symptomatic symptoms of colon cancer. In Belgium, George is accepted based on a standard premium while Peter has to pay a higher premium. Still, both basically represent the same mortality risk because they have a similar chance of contracting colon cancer in the future. When I confronted the underwriters with these cases, they responded: 'I know, it is hard to defend. That boundary is hard to define, isn't it?' (Case 1, underwriter K).

Along the same lines, one may ask what this ban implies for the equality principle in insurance. It seems that, paradoxically, the law itself, by thwarting genetic discrimination, undermines this principle. As Lemmens notes:

> Statutes singling out genetic susceptibility as a category, and offering it much wider protection than other similar health conditions, although intended to promote equity in access to social goals, may themselves be ineffective and to some extent even inequitable.
>
> (Lemmens 2000, p. 349)

The 'right to conceal information' thus benefits those who are carriers of a genetic risk. This can result in the undermining of the equality principle.[9]

Furthermore, one should question the tenability of such legal boundaries between genetic and non-genetic information. Through its juridical embargo, the law affords disputable, undeserved *certainty* to the predictive character of genetics. In medicine, however, it is increasingly acknowledged that the distinction between genetic and non-genetic information is quite artificial. Alper and Beckwith for example, indicate how difficult it is to maintain a distinction between genetic and non-genetic factors (Alper and Beckwith, 1998). According to them, many clinical tests may also provide information on the genetic code. In addition, while genetic tests are commonly defined as information derived from a DNA-analysis, there also exist more indirect forms of genetic testing, like genetic information derived from chromosomes, proteins or via routine urine or blood tests.[10] As this indirect genetic testing becomes more commonplace, genetic and traditional medicine will probably be administered along-side it. Thus it might become difficult for physicians or underwriters to differentiate genetic from non-genetic information. Consequently, it is important to ask whether an exclusive legal ban on genetic information will be tenable in the future.

Finally, the fault-based approach of the Belgian law embodies major implications for the insurance industry. This approach may indeed provide a new meaning for justice in insurance: insurers have to calculate higher premiums when it is the applicant's own fault, otherwise insurers are acting unfairly. Differentiation according to a fault-based approach, after all, denounces the actual meaning of 'actuarial fairness' in insurance. In this regard, insurers have always applied involuntary features in their risk differentiation. For example, various premiums are calculated for men and women, but for the time being individuals do not yet have control over their gender distribution (Brockett, MacMinn and Carter 1999, p. 12). The fault-based approach – as institutionalized in Belgian genetics legislation – may affect the concept of a 'pre-existing condition', which is used as the basis for risk classification. This may cause the mere presence of disease to lose its primary legitimization as the basis for risk selection. As one manager explained:

> It is much more acceptable in public opinion that someone who engages in Alpinism pays a double premium for life insurance rather than someone who suffered from a particular disease two years ago. Yet both may well have the same chance of a premature death. But in the latter case, this is a sensitive issue because it pertains to health factors, or fate, while the former case involves a tangible risk you take and for which you bear responsibility.
>
> (Case 1: manager J)

The idea that risk selection has to be based on an own-fault approach was also raised regularly in interviews with representatives from consumer and patient organizations:

> The insurer should not discriminate on age or on where one lives for instance; this, we feel, is not justified. After all, you cannot help it that you are young or living in some place. Nor should one discriminate on the basis of genetic predisposition. But smoking may be taken into account, we feel. This is a moral standpoint. It applies to the fact that smokers themselves take more risks. Also the fact that someone engages in particular sports: one chooses to take specific risks. Then the insurer has reason to ask a raised premium.

In other words, the moral judgements that involve the controllability of diseases as introduced by the Belgian law on genetics appear to have a cascade effect on the general 'disease' concept in insurance. Insurers tend to fix a premium level that is based on whether the disease is controllable or not for those who are already ill and who want to take out insurance. Accordingly, individual responsibility in the form of higher premiums should only be applied in cases of 'self-inflicted' diseases or risk factors. This trend can potentially have major effects on general insurance principles. Thus, genetics serves as a *catalyst* for a debate on the general workings of insurance policy.

Responsibilities

In this chapter, I have shown how the introduction of predictive medicine in the Belgian underwriting practice goes together with a fault-based approach. This in turn has generated a new attribution of responsibilities: while genetic 'risk carriers' are attributed responsibility collectively, lifestyle 'risk takers' are expected to take their responsibility individually. As noted, this approach is the result of a 'genetic essentialism' in Belgian legislation on genetics in the insurance industry. As such, I explained how regulatory initiatives in the context of genetics can determine a certain practice. In this sense, this legislation is accompanied by major, unanticipated side-effects. The legal imposition of a 'wall' between genetic and non-genetic information raises new questions for the general insurance branch in particular. One of the side-effects discussed here is that moral judgements on the controllability of disease, which are institutionalized through law, can have ramifications for the general disease concept in insurance.

My argument shows that it is questionable whether legislative initiatives can provide adequate solutions to the issue of genetics. It even appears that such initiatives, instead of solving particular problems, may actually cause new ones. This becomes visible when we consider the feasibility of the transformation of socio-normative judgements into legal distinctions or definitions. What does it mean to have control over one's own health? And how do we define the dividing lines between risks that must be carried and risks that are actively sought out? In this respect, the further development of genetics may confront us with new surprises. It may put additional emphasis on lifestyle and individual responsibility. Current developments in behavioural genetics, for example, suggest that genetic factors are partially the cause of many behavioural traits and psychiatric diseases.[11] If one assumes that genetic factors contribute to disorders such as alcoholism (see, e.g., Carmelli, Heath and Robinette 1993) or nicotine addiction (e.g., Straub, Sullivan et al. 1999), the question arises whether people who smoke or have alcohol problems are in fact responsible for the health effects of their behaviour. Ironically, today smoking and alcoholism are presented as the primary examples of bad lifestyles. But this view can also be reversed: The discovery and 'spread' of genetic disease could actually strengthen the notions of lifestyle and individual responsibility. Consider, for instance, a genetic mutation for smoking. Recent studies have aimed at establishing genetic factors for smoking, nicotine-dependency and the inability to quit smoking. This implies not only that the *involuntary* smoker – the gene carrier – is seen as a victim, but also that in the case of those who smoke and who are not gene carriers the issue of responsibility for their behaviour is likely to be highlighted. Detection of genetic components of lifestyle diseases, then, causes the voluntary aspect of non-genetic lifestyle diseases to be emphasized even more.

Paradoxically, the main effect of developments in the field of genetics may well be that individual lifestyle will continue to grow more important regarding issues of health and disease. This becomes even more plausible when we take into account the notion that most diseases are *multifactorial*, meaning that they originate in a complex interplay of lifestyle factors, genetic factors and environmental factors. For example, one may contract specific diseases that have a genetic component largely on the basis of certain types of behaviour, meaning that these individual types of behaviour can be a major factor in the actual occurrence of a genetic disease (McGleenan 2001, p. 41). Instead of assuming a 'genetic determinism' for carriers with a predisposition to some diseases, it seems more plausible, therefore, that we are (all) subjected to various susceptibility levels in relation to disease. In this context, Rose speaks of 'the susceptible self' (2002). It implies that if a genetic mutation in individual cases is discovered for some disease, individual lifestyle habits and preventive measures regarding this predisposition may be pursued in a more sustained manner. Predictive medicine with its early detection of risk factors, then, might help encourage people to deal with their predicted risks properly. In this sense, predictive medicine differs from our traditional way of looking at disease by shifting the focus to the fact that we are the main source of our own health – that even if we do not all draw the same health card there is plenty we can do about it. According to Ewald and Moreau:

> With the arrival of predictive medicine, disease can no longer happen to us blindly. It uncovers a predisposition, a domain that we used to consider as more or less predestined. It affords all of us the possibility – at least in theory – to know our own predispositions. Where disease used to be considered fatal, it now becomes an issue of individual destiny. Undoubtedly, this will significantly change our political anthropology regarding health and disease ... Being ill is no longer something one goes through; it becomes instead a moral risk, which depends on our own conduct.
>
> (Ewald and Moreau 1994, pp. 115–16; my translation)

The risk factors we are aware of make us more responsible for controlling the state of our own health. Thus we are to some degree urged to join 'the worried well' (Harris 1994, in Davies 1998, p. 149) or 'patients *en vie*' (Rouvroy 2000, p. 595). According to Petersen and Lupton, it is possible that those who are seen as 'at risk' and who take no preventive action in this respect are also automatically perceived as failing in their duties as citizens (1996, p. 56). This same approach is currently being developed in Belgium's insurance practice. Individual responsibility for managing one's risk factors becomes the gold standard for assessing one's *fitness* for membership in the insurance pool. In this way, insurers increasingly assess insurance applicants with respect to *decent citizenship* as the basis for their ultimate insurability.

These developments present society with some important challenges. Aside from questions on the desirability of a fault-based approach as access criterion for insurance, it also seems hard to define how someone's control over health or individual responsibility is to be determined. Where, in this matter, do we locate fate, blame or bad luck? Who should or should not be held responsible for their health? For example, the US debate about the risk classification of women who have suffered repeated domestic violence might highlight this issue. The victims of domestic violence were evidently substandard risks from the perspective of the life insurer. Though it was also argued that charging them increased premiums was unfair because the increased risk was beyond their control. One suggestion, then, was that victims of domestic violence should be insured at ordinary rates but that the difference between the ordinary rate and the substandard premium should be passed along to the abuser (Dicke 1999). In other words, financial accountability was redistributed here to the 'cause' of the 'uncontrollable' risk.

It is of little surprise, then, that penetrating socio-political issues arise in these highly plastic risk debates about who should and who should not carry the burden of blame. Moreover, implicit within such choices are quite formidable re-workings of collective and individual responsibility. Predictive medical innovation constitutes new ground in the old debates about individual control, responsibility and blame for health. Finally, this goes to the heart of the basis for *good citizenship* and how this articulates membership – and indeed non-membership – in the insurance pool.

Acknowledgements

This study was funded by a grant from the Research Foundation – Flanders (FWO). A more extensive account of this research can be found in Van Hoyweghen (2006). I would like to thank all the insurance people involved, who allowed me to observe their daily professional activities.

7
Work, Health and Genetics: Problems of Regulation in a Changing Society

Ruth Benschop and Gerard de Vries

Genetic screening at work: a new public concern

At first sight, early detection of a genetically determined predisposition to occupational diseases may appear to be quite an attractive idea. After all, it would provide timely information for individuals with a particular genetic susceptibility to diseases about the risks they will run by engaging in their work. Individuals with a high allergy risk for substances used in a particular industry would be ill-advised to look for a job in that sector of the economy, while a mechanic who is genetically predisposed to carpal tunnel syndrome will be better off with other work. Early detection could be advantageous to employers as well, as testing future employees before hiring may reduce illness and occupational disability related costs.

If this sounds too good to be true, certainly problems soon come to light upon closer consideration. Large-scale implementation of genetic testing could lead to new forms of injustice, exclusion and discrimination. Both in the United States and in Europe, authoritative, government-related institutions have therefore argued in favour of legislation that curtails the use of genetic testing in the workplace.

The various implications of the introduction of genetic techniques in the workplace were studied from quite an early stage. In 1983 the American Office of Technology Assessment (OTA) published a sizable report entitled *The Role of Genetic Testing in the Prevention of Occupational Disease* (OTA 1983). This report introduced the distinction between *genetic screening* and *genetic monitoring* widely used in later discussions: 'Genetic screening involves examining individuals for certain inherited genetic traits. Genetic monitoring involves examining individuals periodically for environmentally induced changes in their genetic material' (OTA 1983, p. 23). Genetic *screening* primarily pertains to the selection of job applicants; genetic *monitoring* applies to checking the health of employees. In this chapter we limit our discussion to the first use of genetic testing.

The preface of the bulky OTA report outlines the occasion for the assessment provided by the report. Information that is obtained by genetic screening could be deployed for prevention purposes, but 'some people fear it could result in workers being unfairly excluded from jobs' (OTA 1983, p. iii). Thus the introduction of genetic screening on the job could lead to discrimination and new forms of social inequality. The report ends with a set of recommendations to the US Congress, advising it to take measures 'to balance the competing interests and to make the value judgments necessary in order to maximize the technology's potential and minimize its risks' (OTA 1983, p. 167)

Early attention to the possibilities and objections regarding the use of hereditary testing in the workplace is also reflected in an advisory report published in 1989 by the Health Council of the Netherlands, entitled *Heredity: Society and Science* (Health Council of the Netherlands 1989). This report concentrates on genetic screening, suggesting that hereditary testing should 'focus only on possible damage to health caused directly by doing the work in question, or on future medical conditions which could affect job performance so as to create risks for other people' (Health Council 1989, p. 165). Other countries have drawn similar conclusions on this issue (European Group on Ethics in Science and New Technologies 2000; Human Genetics Advisory Commission, 1999).

In this chapter we consider these two reports in more detail. We will focus on their analysis rather than their conclusions. These analyses, we argue, are a form of exploration of the future and tend to follow a specific pattern. They are based on assumptions about technology, about the context in which genetic tests will be introduced and about views on the role of government in guiding or adjusting the projected developments. They lead to conclusions about the way in which responsibilities in the area of work and health ought to be distributed. But, we will argue, technology and society are more complex phenomena than these reports allow. Moreover, the way in which they frame the role of government raises questions. The reports' analyses lead to regulation that deserves praise as to its intentions. When implemented, these regulations may increasingly and unintentionally lead to an emphasis on individual responsibility for health. There are, however, few means available for fashioning that responsibility. Moreover, by displacing the problems to the level of individual choice, the possibilities for social learning about the ramifications of genetic testing at work decrease.

From advice to legislative regulation

The analysis performed by the Office of Technology Assessment in 1983 starts with a consideration of the available scientific facts. None of the genetic tests reviewed by the OTA, so it concluded, 'meets established scientific criteria for routine use in an occupational setting.' But 'there is enough suggestive evidence to merit further research' (OTA 1983, pp. 9–10). With this observation, the OTA drops the matter of how new and better tests should be developed, and which standards they ought to meet is left to the world of science. The OTA report is mainly geared to the context in which such tests are used. At the time of the report's publication, this usage was still modest. In a limited study conducted in 1982, it turned out that 6.2 per cent of the companies reported having made use of genetic tests in the period after 1970, while 16.1 per cent expected to do so in the future. A later, more broadly conceived study of medical monitoring and screening on the job concludes that 'little genetic monitoring and screening is currently being conducted by employers' (OTA 1990, p. 7) This study also showed that the large majority of businesses acknowledged the potential ethical and juridical problems of genetic screening.

In 1983 the problem considered by the OTA was largely a problem of the future. To explore the concerns that are expected to rise in the decades to come, the report focused on three main aspects: the economic, juridical and ethical problems they expected to be connected to the use of genetic testing. As to the economic aspects, the authors concluded that because of the large uncertainties about the reliability of the technology, standard techniques for estimating cost-benefit relations and cost-effectiveness could not be seriously applied (OTA 1983, p. 160).

The OTA is more explicit about the various ethical and juridical aspects. First, it explored what effects genetic testing would have on the employer's legal duty to ensure safety in the workplace. The report articulates the concern that the possibility of identifying and excluding employees with a strongly raised risk would lead to a reduction of attention to employee safety in general. In addition, it considers what extra legislation should be introduced to protect employees by addressing the right to a safe workplace, the regulation of medical tests on the job, protection against discrimination, the right to refuse to undergo tests and the right to privacy.

For the analysis of ethical aspects, the OTA appeals to several general ethical principles, with reference to how medical research involving human subjects is organized. The *informed consent* procedures, which had become the standard in that area, aim to guarantee that subjects are well informed, allowing them to make a sound decision when asked to participate in a medical experiment. The OTA argues for a corresponding regulation as to the use of genetic testing in the workplace. By drawing this comparison with procedures commonly used in medical-experimental research, the importance of reliable information and its relation to freedom of choice are stressed. This subsequently leads to a clear recommendation. Where the correlation between test result and disease is limited, the employer is not allowed to go against the interests of the employee, such as refusing him a job on the basis of this test. When this correlation becomes more persuasive, the situation is more complicated. The OTA writes:

> In the hypothetical case of a high correlation between genetic endpoints and disease, the morally correct course of action is significantly less clear. An employer may be justified in allowing a susceptible person to assume the risks on the basis of informed consent. On the other hand, the most ethically feasible course of action for an employer once genetic monitoring identifies a group at increased risk would be to inform the workers and to reduce workplace exposure. Failure to do so would be inflicting harm, and it is unlikely that the group would consent to assuming this risk.
>
> (OTA 1983, p. 148)

In this situation, then, employers have to make choices and they have to inform employees, while employees have to face the question to what extent they are willing to take risks. The OTA assumes the answer to be negative, which seems quite optimistic. After all, not everyone facing the choice between either becoming unemployed or accepting a job that implies a health risk might be able or prepared to opt for the first. In the OTA's analysis, then, the way in which the responsibilities are distributed depends on the reliability of the available tests. But the primary responsibility for health in the workplace remains with the employers. When more reliable tests become available, however, the balance will shift to the responsibilities of employees. The OTA assumes that where adequate information is available such responsibility can also be taken in an informed manner.

The OTA report ends with a list of recommendations to Congress: restrict hiring on the basis of genetic tests; prohibit the use of genetic information, except where the employer can convincingly prove its predictive power; ensure that testing of employees meets the same standards as scientific research with human subjects; and demand full disclosure to employees. These proposals share a single common denominator: they either restrict the employer's room to manoevre or they strengthen the rights of the employee. For employees the availability of sound, scientific, reliable information and individual choice are of crucial importance. The basic view is that when these conditions are met, individual employees will be able to give shape to their right to a healthy workplace in a responsible way.

In short, the OTA submits a proposal for a distribution of responsibilities between employers, employees and government. Genetic testing on the job is not deemed inherently problematic. But specific measures need to be taken in order to guarantee that correct *application* becomes possible. General, legislative measures are proposed to regulate the use of tests properly. The employers retain the primary responsibility for health in the workplace. The OTA signals that nevertheless, in the future, conflicts between the various employee rights may surface. Careful information and individual choice have to ensure that employees really can choose.

The advice of the Health Council of the Netherlands as formulated in its 1989 report *Heredity: Society and Science* basically follows the same line of reasoning, but adds several critical comments. Addressing the role of medical examinations in job applications, the Council rejects 'testing as part of job-selection procedures' (Health Council 1989, p. 19), except in sharply defined circumstances in which the individual's free choice is ensured. Medical examination, the Council feels, should not be used as an instrument to serve, for instance, the employer's financial interests. According to this advisory report – as well as Dutch law – health risks have to be addressed in the workplace, not through the selection of individual employees:

> The use of genetic testing as a selection tool could result in exclusion of workers sensitive to particular external factors, rather than leading to reduced risks and improved working conditions for all.
>
> (Health Council 1989, p. 143)

The Council's report warns against situations in which employers may be tempted to devote less attention to safety issues because employees with an increased susceptibility for certain substances have already been excluded in advance. Such a shortcut would be disadvantageous not only to someone who has to be examined before being hired, but to all employees. It should not be possible for employers to somehow dodge their responsibility to provide a safe and healthy workplace.

The Health Council of the Netherlands emphasizes the significance of several aspects in particular. It indicates that the health with which the employer should be concerned applies to the employee's *current job* only, while his general or future health ought not to be taken into consideration. Furthermore, the Health Council points out, the issue of 'privacy' is particularly at stake in testing (Health Council 1989, pp. 145, 147). Primarily, people's health *on the job* should be at issue in medical examinations, while each crossing into the sphere outside the workplace should be regulated and demarcated. The private domain inalienably belongs to the employee.

In summary, the Health Council disapproves of the use of genetic testing in job hiring: (1) if such tests are geared to the individual's health as such rather than his health as specifically related to work; (2) if they apply to the individual's future state of health rather than his current health; (3) if they pertain to the candidate's private sphere rather than to work only; (4) if individuals are coerced into testing instead of opting for it on the basis of free choice and proper information; and (5) if the usage of genetic tests leads to dwindling attention to the promotion of a safe and healthy workplace.

In the Health Council's recommendations, then, a careful boundary is drawn between reliable medical *knowledge* and its *use*. In addition, the Council emphasizes the distinction between *work* and *private life*. The employer should be responsible for the job environment and the associated health issues. All else is entrusted to the employee in terms of the right to privacy and freedom of choice. Like the OTA report, the Health Council's report assumes sharply delineated boundaries between employer responsibilities and employee responsibilities.

Clearly, the Health Council's advisory report puts the protection of the employee centre-stage. On this point, its advice goes against the trends of the period. In fact, the idea that the employee is in need of protection has increasingly been replaced by a style of thinking that starts from the employee's individual responsibility:

> The employee is no longer considered an object to be protected, but an acting subject who is thought to be capable of developing responsible health behaviour. Subsequently, the employer is called on to exercise his ability to influence this behaviour via policies aimed at prevention and sanctions.
>
> (Trommel and Van der Veen 1999, p. 104)

The 1989 Health Council advisory report tried to limit the application of this emerging style of thinking by linking genetic tests to an individual's *current* job. The same approach would inform the Dutch Medical Examination Act (Wet Medische Keuringen) adopted in 1997. Also at the policy level, the Health Council opts for a subtle balance between individual responsibility and legal, government-imposed obligations. Specifically, practice itself should be allowed space for self-regulation to effect adherence to the conditions and views articulated by the Council. It also suggests a specific time limit, though, after which legislation will become necessary.

The arguments by the Health Council of the Netherlands and OTA, in part at least, were indeed put down in legislation. The 1997 Dutch Medical Examination Act stringently regulates medical examinations in the context of job applications. Earlier agreement on this had been reached in the *Protocol on Job Hiring Examinations* (*Protocol Aanstellingskeuringen*, 1995) to which many agencies subscribed, including the KNMG (Royal Dutch Medical Association), unions, and occupational safety and health services organizations. Today medical examinations as part of job application procedures are allowed only for positions in which special medical requirements apply, and in such cases the examination has to meet special legal conditions. For instance, the employer first has to ask for advice from the Occupational Safety and Health Service on the content and legality of the examination, and it has to be clear in advance which medical requirements belong to a specific job and how it is to be tested whether the applicant meets them. From the start it must also be clear to the applicant that such examination will be part of the application procedure. The *Protocol* even explicitly articulates that neither questions on, nor testing of, hereditary predisposition is allowed to be part of an examination in the context of a job application.

With the Medical Examination Act, the potential dangers formulated by OTA in 1983 and the Health Council in 1989 as to the usage of genetic testing in job hiring examinations seemed to be averted for the time being. If this legislation did not fully exclude the usage of genetic testing, it became tied to strict conditions in order to prevent its improper deployment.

Yet one should be hesitant to draw firm conclusions at this point. In 1998 the American Equal Employment Opportunity Commission, relying on several studies, reported both on cases of genetic discrimination and on people who do not want to make use of genetic tests – or want to keep their results secret – out of fear that their results will be detrimental in obtaining work and taking out insurance (Department of Labour et al. 1998). In 2003 the European Group on Ethics in Science and New Technologies reported that it is still unclear how many employers make use of genetic testing (European Group on Ethics in Science and New Technologies 2003). This working group assumes that its use is generally more widespread in the United States than in Europe. It also claims that it is logical that where a technology is still hardly available, its usage will be limited. In England one case of such usage was known, while research there revealed that several CEOs would approve future usage under certain conditions. Similarly, the evaluation of the Dutch Medical Examination Act seems to suggest that the institutional pressure to obtain increasing information on the health of future employees has not diminished. Although the number of job hiring examinations dropped, 46 per cent of the companies try to gain 'information on the health of an applicant' in other ways. Moreover, people are often examined 'without justification of the job description involved'. Finally, many company doctors save examination records of those who are appointed without their permission (Borst and Hoogervorst 2001).

If the threats that originally motivated the OTA and the Health Council of the Netherlands to formulate their recommendations seem to be in part averted, the problem has not entirely disappeared. This is why the specific analyses provided at the time – which in the Netherlands resulted in legislation – still deserve closer scrutiny.

The limitations of overly rationalistic technology analyses

When in 1983 OTA and in 1989 the Health Council concerned themselves with the problems of genetic screening and work, the genetic testing technology was still in its infancy. It was anticipated at that point, though, that this technology would improve in due course and that the validity and reliability of tests would increase. At issue in these reports are problems of the future. This future, however, is conceived as more or less given. This expectation is based on a specific developmental model of technology that assumes technologies to run through a particular course, often represented graphically in the shape of an S-curve: after a slow beginning the development of a technology will accelerate until it has reached its 'mature' state.

Sociologists of technology, relying on historical studies of technology, have questioned the self-evidence of this way of thinking about technological development (Bijker 1995; Latour 1996). The assumed development pattern is certainly no automatism. Its popularity relies heavily on looking back on the histories of successful technologies. For these technologies, it is usually not hard to identify the point at which they came into being and to subsequently trace the path of their further evolvement.

From this angle, one is likely to believe that a technology's 'mature' use was already built into its original design. But hindsight is an unreliable tool for a historian. A more critical tracking of technology development shows a more complicated picture. Often, the initial promises of a technology are not realized. In due time variants emerge that do not make it, while new and unexpected applications may gradually present themselves. In the context of technology development, organic models of development that suppose that later developments are encapsulated in the seed have to be viewed sceptically. An encyclopaedia may be filled with descriptions of technologies that did *not* work as initially conceived or that received an altogether different application than first anticipated. That we merely have to cast a glance into a laboratory to see society's future is a popular misconception. An important part of technology development takes place long after technologies have left the laboratory. In their actual usage and interaction with the social environment, they take on a different shape. When cell phones were introduced, no one had foreseen that this technology would become popular, in particular among youngsters, and that text messages would play such an important role.

Analyses of future social consequences of new technologies, then, have to be considered with suspicion. Apart from the arguments advanced in sociology of technology studies, in the case of genetic screening in the workplace there are also convincing biological and epidemiological reasons for doubts about the initial promise of this technology.

Evidently, over the past two decades major progress has been made in the field of clinical genetics. However, it is wrong to extrapolate these advances to screening in relation to work. First, the accomplishments of clinical genetics mainly apply to diseases that are linked to one gene or a very limited number of genes. But work-related genetic disorders, such as cancer and allergies, are tied to the interactions of various genes, as well as between genes and their environment. Predictions on polygenetic diseases are not only inordinately more complicated in quantitative terms than predictions about monogenetic disorders; they are also quite different in qualitative terms. While predictions on monogenetic disorders have a high level of certainty (carriers will die from the disorder, unless they die before of another cause), more complex cases as a rule involve predictions with a fairly low probability. Second, making predictions about diseases outside the context of families with a known anamnesis – and without the possibility of studying the genetic constitution and anamnesis of family members – calls for another methodology than clinical-genetic testing. The prevalence of the trait studied in the general population then becomes a major factor for a test's maximal predictive power. In rare diseases the tiniest error in the sensitivity or specificity will strongly decrease the accuracy of the results. Even tests with high distinctive powers cannot do otherwise but produce poor predictions. A simple calculation demonstrates that if, for example, the trait studied occurs in 0.1 per cent of the population and the test has a sensitivity and specificity of 99 per cent, only 9 per cent of those whom the test identifies as a carrier will also in fact be a carrier.[1] Nine out of ten people who the test identifies as carriers are not and hence are rejected erroneously. Apart from the obvious injustice that is done to those who have been tested, employers relying on such a test in hiring staff would thus exclude many more people than intended. Especially in a tight job market, they can hardly afford to follow this wasteful path.

This situation may be changing. Thanks to DNA-chip technology, it is now possible to study the expression of a whole series of genes in a short time, as is true of gene combinations. This may significantly increase the sensitivity and specificity of tests. For the time being, however, this only solves part of the problems in the study of the genetic aspects of susceptibility to work/environment-related disorders. There is still the question of how many carriers will actually fall ill when exposed to specific working conditions. More epidemiological study will be necessary in which accurate data have to be gathered on the exposure history of each individual in the sample's population. To achieve scientifically relevant results, especially when a large number of genes and gene combinations is to be studied, it will be necessary to study large sample populations. As the number of individuals in a sample rises, however, it becomes ever harder to gather accurate data on the degree of exposure. As a result, notably the study of diseases with a late onset, such as work-related cancers, will probably continue to run up against severe limitations. In this context, a commission that in 1999 advised the British government on the use of genetic testing concluded that 'it will take major developments both in our understanding of common diseases and in genetic testing itself before genetic testing becomes a serious issue for employment practice'(Human Genetics Advisory Commission 1999)

That genetic testing has barely made inroads in the workplace, then, should perhaps be explained by these technical limitations regarding its usability rather than by the legal measures taken. This is not to say that such measures have become meaningless. The *illusion* that genetic tests supply accurate predictions about the health condition of applicants, including their susceptibility to disorders, and the ensuing *temptation* to use such tests constitute a no less real danger than their actual use. Legal measures as articulated in the Dutch Medical Examination Act, which force an employer to report the use of tests in advance to the Occupational Safety and Health Service so as to prevent improper use, are one way of dealing with such temptation. However, as indicated, actual social practice deviates from what the legislator had in mind on several points. If we are seduced by the illusions of predictive tests, let us not be seduced by the illusions of predictable laws as well.

Regulation in a changing society

Not just the assumptions about the technology involved deserve attention. The analyses by the OTA and the Health Council of the Netherlands also start from assumptions about the context in which the technologies under review will be applied and which only in part cover current employer/employee relationships.

This may be illustrated by the OTA report. It suggests that genetic testing would be useful, in particular, in the case of exposure to chemical substances and ionizing radiation (OTA 1983, p. 23). Although people have long been exposed to chemicals and radiation, especially in some professions, the risks have substantially gone up since the Industrial Revolution (OTA 1983, p. 67). Thus the OTA decidedly situates genetic testing in an *industrial* context, as evidenced by its further elaboration of the parties that are active in and around the industrial workplace. The OTA report claims that businesses have different methods for dealing with health risks on the job, and genetic testing could just become part of these methods (OTA 1983, p. 27). What the report's authors have in mind is a fairly large company with numerous employees. This becomes clear when closer attention is devoted to the relationships between employer and employees, and reference is made to company doctors as well as unions as possible representatives of employees when juridical questions are at issue. The assumed infrastructure is that of the industrial branches of business.

Similarly, there is no lack of clarity on the body that can authoritatively enact the proposed regulations: the OTA aims its report exclusively at Congress, which is responsible for taking general legislative measures in the United States. In this respect there is no mention of a possible role for other parties, such as private insurers and professional associations. Congress is supposed to be the institution to deal with OTA's concerns. But whether the proposed policies can effectively be implemented in practice remains unaddressed. It is assumed that those who are charged with the management of public life – government and parliament – also have sufficient insight into society to turn planned policies into successful policies.

The Health Council report is less specific as to situating the problems involved. For both hiring procedures and the health monitoring of employees, the report assumes the availability of medical services (today an internal or external occupational safety and health service) whose task it is to take care of the medical guidance of employees during their employment. It should be pointed out that in its report the Health Council is hardly explicit about the nature of the workplace. Although its report may refer to employees and employers, frequently more general terms are used, such as 'the person involved', 'the person tested', 'the job situation' and 'the job environment'. Where the OTA report clearly situates the application of genetic tests in an industrial context, the Health Council's is more general. Still, the prevailing image suggested by its report is similar: an employer with numerous employees who has a service available for monitoring the health of his employees, while at the same time an institutionalized selection moment is also assumed. In the work situation, clear interests and power inequalities – such as 'the dependent position of the person tested' – are present and there is the supposition of transparent and effective regulations aimed at protecting the employees. In contrast to the OTA, however, the Health Council explicitly recognizes that regarding the new policy on genetic testing, experience first needs to be gained before effective legislative measures can be taken. The Council proposes offering the parties involved the opportunity to develop decent procedures on their own initiative. Ultimately, however, the Health Council too feels that legislative regulation is necessary.

The recommendations by the OTA and the Health Council, then, are based not only on a specific image of the promise and further elaboration of the techniques involved in genetic screening, but also on a specific image of the situation in which such tests will be used, namely an environment of already explicitly and well-regulated work relationships. Of course, this comes as no surprise, considering the addressee of the recommendations provided – government and parliament – and their objective, which is to come to the drafting and enactment of legislative measures and regulation.

But how representative is the assumed situation for the current employment realities? The OTA report discusses genetic testing in a context that to a large extent is by now out of date. In spite of the fact that television programmes that seek to represent work in general terms invariably show spectacular images of port activity, iron works, an automobile factory or a chemical plant, the view that the centre of economic gravity is in the heavy and dangerous industries should by now be relegated to the domain of folklore. The majority of people in the prosperous West nowadays earn their living in occupations such as processing information, providing services and engaging in meetings and deliberations.

On this point the Health Council report, which sketches a more generic work situation, is immediately more realistic than the OTA report. Yet the notion of stable and well-ordered work relationships that also informs the Council's report has been subject to eroding. In the past decade, we have witnessed the emergence of new forms of employment relationships. Examples include part-time work, temporary work, stand-by and substitute contracts and outplacement constructions. The number of people with a flexible job has risen by 60 per cent. In addition, an increasing number of people opt for being self-employed. The standard job – five days a week, from nine to five, from one's adolescent years to retirement – is becoming a rarity. Increasingly, individuals, often after having functioned as an employee for some time, opt for a life of being 'self-employed without personnel'. For this growing group, the traditional roles of employer and employee coincide, as it were. As such, the reports by OTA and the Health Council, which advocate accurate regulation and distribution of the responsibilities of employers and employees for health in the workplace, hardly apply to the self-employed. Rather than hiring employees, they accept specific assignments or work with specific kinds of clients.

Self-employed individuals, like all others who work, operate within the legal frame of rules as to how their safety and health must be ensured. This frame, however, is different from what counts for traditional employers and employees. Government, too, acknowledges that the issue of occupational health and safety in the case of the self-employed (and, by extension, small and medium-sized businesses) is far from a simple issue. Businesses in this sector are so small that most do not have the financial leeway to address occupational health and safety in a systematic way. This in part explains why they are much harder to reach by government agencies in charge of occupational health and safety policies. Unlike the traditional employees and employers to whom proposed regulations apply, the autonomously operating self-employed have no fixed relationship with the Occupational Safety and Health Service and its instruments for monitoring work and health and reintegration trajectories and such. Arguments valid to others are irrelevant here. If the Health Council stresses the distinction between the sphere of work and the private world, for the self-employed the two spheres frequently fuse, notably if they also work at home. Many self-employed appreciate their work situation precisely for this reason and for the freedom it provides to fashion their work on their own. In many cases they seem to take lack of protection from the government for granted (Evers 2000).

That general legislative protective measures do not apply to each and every person is of course not a new phenomenon, nor is it specific to the domain of genetics and work. In innumerable areas governments are confronted with the situation that generally formulated measures are poorly tailored to a society that can increasingly be seen as flexible and differentiated. This is why various authors and advisory bodies have endorsed guidance models that take our modern culture of difference into better account; they also underscore that, to public administrators, society in many respects is an 'unknown' (Van Gunsteren and Van Ruyven 1995; RMO 2002). Likewise, some argue for flexible and adaptive systems which at the implementation level mainly rely on feedback (and this would also apply to problems of safety and protection of citizens) (Mertens 2003). Familiar answers to this situation are the introduction of free market structures, enabling legislation, relatively autonomous executive bodies and self-organization of professional monitoring. At the same time, however, it must be acknowledged that where flexibility and differentiation are chosen as starting points, government will soon run up against the boundaries that come with the core principles of the constitutional state, such as equality before the law, and the role attributed to government in regard to organizing social solidarity.

Genetics and work as a challenge to public governance

To protect employees against foreseen negative effects of the introduction of genetic screening in the workplace, in most if not all industrial nations governments have taken legislative measures that sharply curtail the use of genetic testing in medical examinations in the context of employment procedures.

Preceding this legislation are specific analyses that are grounded in assumptions about the development of genetic technology and the specific job situation in which it will be applied. As argued above, this invites several comments. For a growing number of employed, the moment of accepting work is no longer tied to a formal hiring procedure. Increasingly, the organization of work also deviates from the institutionally arranged interaction of employers and employees which is assumed by these analyses. Moreover, it is possible to criticize the expectations concerning technology development that play a role in these analyses.

That the preconceptions regarding technology development formulated 20 years ago proved not to be correct as of yet does not, as pointed out already, render the regulations involved meaningless. Employees also need to be protected against the *illusion* that genetic screening produces meaningful results. And although for now there are good scientific reasons for sustained scepticism about the possibilities involved, new scientific or technological developments may change this situation. Solid legislation in the area of job hiring examinations may then prove to be a major social achievement.

Yet this legislation, and the logic that engendered it, may unintentionally have negative effects. The legislation relies on the notion that knowledge leads to just and responsible decisions. This notion reverberates at two levels. First, it resonates in the model of analysis and guidance which is mobilized: in the assumption that a responsible government, before it applies regulation, is cognizant of the analyses made by organizations such as the OTA and the Health Council. Second, this notion is grounded in the recommendation that, when reliable and valid tests become available, individual employees ultimately decide on their application, and this means they need to have the relevant information at their disposal on the risks they, as the carrier of a genetic predisposition, will run in a specific job context.

Public administration experts have articulated substantial doubts about the tenability of the model of analysis and guidance that constitutes the basis for the role of government as indicated (Van Gunsteren and Van Ruyven 1995). Of course, these authors do not deny that in the public sector administrators also have to rely on their common sense. Yet what is challenged is the assumption that administrators can really 'know' society. Those who start from common sense would do well to assume that in many respects society is 'unknown'; that despite all analyses and predictions, there will *always* be surprises. What should serve as a starting point is not the certainty of knowledge, but the awareness that we are living in an uncertain world. In the area of genetic screening on the job, at least two such uncertainties can be indicated: we simply do not know how this technology will develop, nor can we but speculate about the future development of the organization of work. Those formulating regulations will therefore have to reckon with such uncertainties. In legislation they will have to take into account that innovation, flexibility and differentiation are normal social phenomena. The consequence is that legislation should allow for explicit arrangements aimed at enabling social learning processes. The existing legislation on genetic testing as part of medical examinations in job hiring hardly reckons with this insight derived from public administration theory.

At an individual level, a similar situation presents itself. In the analyses of both the OTA and – albeit to a lesser degree – the Health Council, it is assumed that individual employees, if they have reliable information at their disposal, can make choices in responsible ways. Where the protection paradigm is replaced by the model in which the individual responsibility of the employee as acting subject comes first, this notion is even explicitly embraced. Employees are then seen as individuals who 'manage' their own life, also in the workplace, and who make choices, opt for a lifestyle and actively understand and weigh the risks involved (De Vries 2003, 2005). Taking responsibility is presented as a natural aspect of emancipation and of the highly valued notion of individual choice.

But does this argument stand up to scrutiny? And is individual choice emphasized because more and more people have become convinced of the liberalist doctrine? In many respects Foucault's observation that in our time, liberalism is not so much a conviction or ideology but a form of governmental rationality seems more to the point (Foucault 1994, vol. IV, p. 36). The emphasis on individual responsibility, apart from being a sign welcomed by many that people are finally being taken seriously, also constitutes a solution for the managerial problem that society is *unknown* in many respects and that because of innovation, flexibilization and differentiation, general measures frequently miss their target: instead of solving problems on a managerial or administrative level, then, they are transferred to decisions at an individual level.

However, first, it is doubtful, especially where it involves job situations, whether people have the means to realize the responsibility handed to them. Do employees who have reliable insight into the risks they run in a given workplace really have something to choose? Not only the genes with which we are born are in many respects a matter of destiny, but also our opportunities in the job market. The assumed freedom to choose is not really available to many. The Health Council rightly acknowledged this in its 1989 advisory report by insisting – against the trends of the time – on legally regulated protection of employees. It is a question, though, how long the echo of its insistence will resound once the above-described technological developments and changes in the job market materialize.

The second comment concerns the loss of social learning power that emerges when problems of public administration are transferred to choices at an individual level. Where regulations exist that are linked to well-functioning executive bodies, one may expect that problems of regulation that become visible in practice are identified at some point. Where they are treated individually, it is not guaranteed that such problems will become *public*. Those involved will easily think that they are on their own, coping with individual problems. Where choices are made at an individual level, therefore, new institutions are needed that bring potential negative consequences of developments into view *as public problems* and allow them to be addressed as such (Nelis, De Vries and Hagendijk 2004). In the field of genetics and work, however, such institutions are not yet available. The developments in this field, then, present our society not only with ethical dilemmas and issues of solidarity, but also, and perhaps first and foremost, with a challenge to public governance.

8
Genetic Risks and Justice in the Workplace: The End of the Protection Paradigm?

Rein Vos

New tests, new responsibilities

In recent years different groups and organizations have made a plea for the restriction of the opportunities for genetic testing of the susceptibility to occupational diseases of incoming employees. These pleas have led to laws and regulations which aspire to prevent new forms of discrimination and social exclusion, and express the undesirability of lifting the obligations of employers to maintain a healthy work environment. Specific and sensitive tests to determine the genetic susceptibility for health risks in the workplace are still not available and one can with good reason be sceptical of the promises that such tests will appear on the market at short notice (see Chapter 7).

Nevertheless, the various normative questions involved need to be raised. Is it really unreasonable, for instance, that people starting their baker training are tested on their susceptibility to developing baker's eczema? Or that future painters or upholsterers are tested on their susceptibility for what is called painter's disease? Why lock the stable door after the horse has bolted? But who has to decide upon these issues? What are the responsibilities of employers, employees and the government with respect to workplace health? To be sure such tests are not available, but it is useful to discuss such normative concerns before these tests come to the market. Moreover, normative questions do not apply only to genetic tests. Other forms of employment examinations or predictive tests in the workplace are at issue as well.

In this chapter I take up these normative issues. I will use the debate on 'painter's disease', also known as Organic Psycho-Syndrome (OPS), but also some other examples, including RSI, burnout and work stress, to study the framing of workplace risks and the responsibilities of dealing with them.[1] From this it would appear that the protection paradigm in occupational health has a restraining influence on social learning processes related to genetics and work and the new responsibilities of the stakeholders. But first, let's take a closer look at the protection paradigm.

The protection paradigm

Concerns about work and health have a long history and are in fact a cornerstone of the welfare state. Since the Industrial Injuries acts of the late nineteenth century, social regulations involving disease and occupational disability constitute the model for the social welfare system of modern Western societies (Baker and Simon 2002). Our thinking about work and health suggests a specific pattern that I would like to identify as the 'protection paradigm'. Employers have a legal obligation to ensure a healthy work environment. In this context, the causes of risks and their prevention have been increasingly emphasized. This shift was articulated in the 1983 Dutch Occupational Health and Safety Act (or Labour Conditions Act or 'Arbo-wet'). In 1999 its revised version took into effect the legal requirement on employers to introduce a permanent cycle of improvement. It applies to identifying and evaluating risks (RI&E), formulating relevant policy plans, and implementing preventive and protective measures. The protection paradigm is geared to causes linked to physical agents such as radiation, noise, dust and chemical substances – also called 'source' – and measures that can be taken to counter them. In the case of hazardous chemical substances this means: (1) eliminating hazardous substances; (2) substituting them for less hazardous or non-hazardous substances; (3) shielding the source; (4) adapting work processes; and (5) supplying personal protection measures. Where complete removal of hazardous substances is impossible, their reduction should be maximized.

The protection paradigm starts from the notion that risks can be objectified, calculated, standardized and influenced. It is assumed that one can measure the exposure to substances, identify the effects on health, establish dose-effect relationships and subsequently determine standards – limit values, mostly expressed as maximally acceptable concentration (MAC) values – that constitute the basis of measures aimed at excluding and preventing risks. This simultaneously defines the responsibilities of the protection paradigm's various parties: the government develops a legislative frame for preventive measures, the details of which may be negotiated by employers and employees; employers implement prevention policies, which the Occupational Health and Safety Act formulates as their 'duty to provide care for' (i.e., the obligation of the employer to take care of all necessary measures to guarantee a safe and healthy workplace); and the employees, finally, appear as the quasi-passive object of those policies and merely have the responsibility not to behave recklessly or carelessly.

This frame serves as the general background against which 'individual' factors or situations are judged,[2] wherein three specific roles stand out: the employer who does not follow up on his or her duty to provide care, or the 'negligent' employer; the employee who does not take on his or her responsibility to use the recommended protection measures, or the 'troublesome' employee; and the employee who should accept his or her occupational disease as an instance of 'fate', or the employee as 'victim'. In the latter case it remains an issue whether or not the destiny involved is just and whether the related costs have to be compensated publicly or privately.

As soon as problems present themselves within this paradigm, the parties involved are pitted against each other. For example, unions may pursue better prevention for their members. In this respect the Dutch Federation of Trade Unions (FNV), the largest trade union in the Netherlands, has set up a specific Bureau for Occupational Diseases to assist employees in submitting claims against employers. But employees who feel victimized may also go to court on an individual basis in order to try to receive compensation for harm suffered. Closer consideration of the debate on painter's disease may illustrate the various problems that thereby arise, and also demonstrate the protection paradigm's actual functioning.

The protection paradigm in practice: painter's disease

'Painter's disease' is the popular name of what is called organic psychosyndrome (OPS) or, in the medical scientific community, 'chronic toxic encephalopathy' (CTE). OPS is a serious form of brain damage related, among other things, to exposure to volatiles in the workplace. Many of these organic substances, such as solvents, have neurotoxic effects. This disease frequently occurs among house painters, hence the name 'painter's disease', but also in many other professional groups, such as paint sprayers, shoemakers, upholsterers, carpet layers, nail stylists and workers in chemical industries.[3] It is caused by poisonous solvents used in paint, glue and cleaning agents. Exposure to these substances may damage the nervous system and lead to concentration disorders, memory loss and personality changes, potentially incapacitating those afflicted permanently. Exposure mainly takes place through respiration, but also via the skin. At first the complaints are insidious, disappearing again during weekends or on vacation. When the exposure persists, the complaints increase and become more serious (Hageman and Van der Hoek et al. 1999; Van der Hoek 1998).

In many ways OPS is hard to deal with. The Bureau for Occupational Diseases, set up by the FNV, provides a lively description of what people with this or other chronic work-related disorders experience:

> Before those who suffer from a job-related disease knock on our door, they pass through terrible ordeals, including lack of understanding from their employer and a company doctor who doesn't quite know what to do. Because they have to reintegrate and most are eager to go back to work, they try to take up their own job again. Often this even causes more damage, especially among RSI patients. Commonly, family physicians do not know what to do with work-related diseases either. Meanwhile you, the ailing individual, are trying to recover. You shop around in the medical arena and often in the alternative circuit as well. Next, you are medically examined for permanent disability benefits. This means that within one hour your financial future is decided. You are ill and your income will drastically go down. Your illness forces you to adapt your lifestyle, which has to be done again when your income starts dropping. This makes it all very hard to give life meaning again. Therefore, your decision to knock on our door is a courageous one. You engage in battle with your former employer to receive financial compensation for a disorder *he* is responsible for.
>
> (Homepage FNV Bureau for Occupational Diseases 2001)

That there are individuals who suffer from this clinical syndrome is evident, but on the exact causes and disease mechanisms little is known. In the world of research and beyond, OPS is also called 'the Scandinavian disease' because Scandinavian researchers reported on this disease early on. In 1976 Denmark was the first country to classify OPS as an occupational disease (Loo and Van Weme 1997). Initially, however, researchers from the Netherlands, Germany and England failed to replicate or verify this disease. The government began to pay attention to OPS in the Netherlands only by the 1990s, in part because one division of the FNV – its timber and construction union – had taken the initiative to establish an OPS patient organization. This organization started an active lobby in Dutch politics, with support from the FNV. Partly as a result of this lobby, several authoritative agencies such as the Health Council of the

Netherlands and the Social and Economic Council of the Netherlands issued advisory reports on OPS and solvents in the 1990s. This has led to a prevention policy aimed at reducing and preventing OPS. Two types of policy measures were initiated. First, it became mandatory for companies to substitute organic solvents or products that contain organic solvents for products that contain no or only few solvents.[4] This has meanwhile been enforced through legislation for specific (indoor) situations in the paint, housing and hardwood floor sectors, the graphics industry, automobile paint shops and automobile damage repair shops. Second, an exposure standard was formulated: the so-called Maximally Acceptable Concentrations (MACS). This standard is aimed at reducing the exposure as much as possible in order to reduce the risk of contracting OPS.

OPS or CTE remains a thorny issue for occupational toxicologists, however. Most suspect that individual sensitivity plays a role. This is why, the Health Council argues, 'timely diagnosis of early CTE symptoms in employees is called for, so as to prevent its further development. This is important, in particular, for persons with increased sensitivity' (Health Council of the Netherlands 1999, p. 2).[5] The probability is that OPS only surfaces in response to peak exposure and only if one is very sensitive to it. Many people come into contact with solvents after all, but the number of them who fall seriously ill is limited. Possibly, employees with a specific neuropsychological constitution are particularly at risk. There is consensus among toxicologists that chronic exposure to concentrations below the MAC value does not raise the chances of contracting CTE. Recently, based on advisory reports by the Health Council and the Social and Economic Council Commission on Working Conditions, standards were formulated for the reduction of so-called peak exposure to organic solvent vapours, even though here too the explanatory biological mechanism for this correlation is absent. In the absence of full certainty about the very existence of OPS, everyone agrees that OPS *ought not* to exist!

The protection paradigm has led to a concrete policy, comprising a clear standard, a substitution obligation and a strict prevention policy. This is not just true in the case of painter's disease. There are countless examples of successful substitution having led to huge reductions in exposure to harmful substances. In the bakery sector a reduction of the risk for baker's eczema was realized by introducing dough that contains little amylase. Exposure to welding fumes in the metallurgic sector is regulated in part through personal protection measures.[6] Since 1998 the Dutch government has deployed so-called 'arbo-covenants' as a policy tool. Arbo-covenants are formal agreements with professional sectors on objectives for reducing exposure to various substances and other risks.

Table 1   Objective Arbo-covenants

Risk	Objective (reduction of number of exposed)	Realization	Risk population (nationwide)	Reduction (nationwide)
Lifting	30%	2005	1,300,000	390,000
Work pressure	10%	2003	1,700,000	170,000
RSI	10%	2001	1,000,000	100,000
Noise	50%	2002	340,000	170,000

Source: Bus 2000.

In 1999 the government negotiated deals with employers and employees in 'high risk' sectors on reducing specific work risks – related to lifting, work pressure, RSI, damaging noise, solvents, allergenic substances and quarts – and formulated explicit goals (see Table 1).

Businesses, however, should not only consider substance substitution and risk reduction as a cost. These measures may also have positive effects for the sectors involved, such as enhancing their public image or professional expertise and boosting product innovation.

In the case of solvents and painter's disease the feasibility of a full ban of products that contain solvents has been extensively discussed (Loo and Van Weme 1997). The large paint and glue manufacturers argued that it was very possible to produce such products against roughly the same cost. Moreover, a study by the Centre for Research and Technological Advice (COT) would also have demonstrated that in the long run the products without solvents have the same results (shine and protection) as those with them. However, because products with no or hardly any solvents are harder to apply in cold and wet weather, they require professionals to use a special technique. These professionals themselves did not necessarily consider this a drawback, though, and even argued that it could cause their business to go up because do-it-yourselfers would be more often inclined to leave it up to professionals to do the job (Loo and Van Weme 1997). If employers and employees were divided on this issue, government and parliament opted for a combination of a considerable substitution obligation and a stringent prevention policy.

Juridicalization

The existing Labour Inspection guidelines are designed to provide 'sufficient protection' against unhealthy situations in the workplace. In actual practice, however, as the Health Council of the Netherlands has claimed, these guidelines appear to be:

> barely effective, given the fact that cases of CTE continue to occur. The commission suspects that those in the workplace do not attach enough direct attention to self-protection or do not or cannot provide enough protection to their employees. There is, apparently, a lack of knowledge, means and independent control of compliance.
>
> (Health Council of the Netherlands 1999, p. 38)

The FNV arrived at the same conclusion.

> In 2002 the Labour Inspection observed that in as much as 42 per cent of the companies that use hazardous substances, protective measures are insufficient if not altogether absent. Agriculture, construction and industry are frontrunners, but this problem also occurs in many hairdressing salons and cafes or restaurants. Particularly in small companies, adequate measures are absent.
>
> (www.bbzfnv.nl, 26/06/2004)

What happens when existing guidelines do not result in effective prevention? Regarding painter's disease there was an extensive effort on the part of the FNV. 'I used to be a number, now I am a nil', was a slogan used at a meeting of 'Friends of the FNV Bureau for Occupational Diseases' (Utrecht, 22 November 2001). It involved a cry for help from an OPS patient, a victim of disobliging employers. The task the FNV sets for itself is clear: helping the victim and pressing charges against the employer. It is 'a matter of decency', so the FNV thinks, that an agency for supporting OPS victims should be established. Such an OPS agency can prevent the double agony for victims of painter's disease, who 'first contract a serious and irreversible disease and next fight an exhausting battle for compensation' (FNV Bouw Nieuws 2003b, pp. 14–16).

In cases of individual employees, the judge eventually has to rule on compensation in court. This road makes heavy going and not everyone is able to do it. A legal case may last as long as three years, if not more, while the process of objectifying damage is lengthy and complicated. How much damage was caused? How long has the disorder been there and how serious is it? Is there any cover from insurance? Does it indeed involve an occupational disease? In the Netherlands employees increasingly press charges against their employers to win financial compensation for harm suffered,[7] whereby Article 7: 658 of the Dutch Civil Code plays a central role. Employees who become permanently or partially unable to work may hold their employer responsible. Where previously employees had to prove that their disease was caused by their work, since 1997 the burden of proof has rested with the employer. This so-called legal 'reversal rule' basically means that a causal relationship between disease and work is assumed if employers cannot prove that they complied with their obligation to provide for their employees.[8] The issue is of course what, exactly, employers are obliged to provide for and what it means to conduct a labour conditions policy 'as effectively as possible', based on 'the current level of science and professional services'. The law requires that the measures employers take are broadly accepted by 'experts' as applicable in practice. These measures may be stipulated as standards in the Labour Conditions Act itself, in Arbo-decisions, as well as in standards in other regulations and publications, such as employment contracts and arbo-covenants.[9]

In the literature and jurisdiction, as well as in the field, one may encounter 'painful' stories that illustrate the extent to which court cases linger. Both sides bring in experts on work issues, medical issues and juridical issues. Insurers and employers may claim that the disease or damage is fully or partly a result of private circumstances or an effect of the employee's personality structure. In the jurisdiction on RSI this argument is often brought in, but it has not been honoured as of yet. It is up to the employer to prove that these private circumstances really occurred and are also the cause of the complaints. Employers who failed to comply with their obligations to provide occupational health and safety measures are liable. However, it is not an all or nothing matter. The employer does not necessarily have to be held completely liable for the harm suffered. It may be a case of proportional liability. In this respect, the Working Group on 'Collective compensation for OPS victims' reported:

> Insurers do their own investigation of the full facts and medical causality. Commonly insurers and their advisors tend to attribute the disorder to private circumstances or personality disorders. In some cases one tries to aim for proportional liability, when there are more factors – aside from working conditions – that can have caused the damage. (Working Group on Collective Compensation for OPS Victims 2003, pp. 10–11)

Often innumerable factors are involved. Many OPS patients have a broad range of mental complaints, including masked depression, coping with bereavement, after-effects of divorce, relational problems and stress.

They have other psychological diagnoses, may suffer from other occupational diseases – back complaints, shoulder and neck complaints, hearing damage – or from other somatic disorders such as heart infarct, brain infarct and bronchial disorders. The specific weight of each of these various factors has to be established in court. No wonder, then, that cases linger for years. To come to collective compensations, the FNV has been arguing for the establishment of an OPS agency. Insurers also favour such regulation, but mainly in terms of *risque professionel*, meaning a system of compensations for occupational disability from a fund based on a legally recognized list of occupational diseases as exists elsewhere in Europe, such as in France and Germany. How this will play out remains to be seen, but it is questionable whether today's protection paradigm will still be as central tomorrow.

Politics of pity

The protection paradigm, which frames the employee as a passive subject, leads to what might be called a 'politics of pity' (Boltanski 1999). Such a politics produces a number of effects. First, it forces a division between those who 'suffer' and those who do not; we do not say, for example, that employers or insurers are suffering; OPS victims are suffering. Second, it focuses the attention on what can be seen: the 'spectacle of suffering'. This means that a politics of pity is not directed towards action, namely the power of the strong over the weak, but towards the stirring observation of the 'unfortunate' by a 'fortunate person'. Third, such politics starts from the view that the unfortunate's bad luck or fate is unfair. After all, who would dare to suggest that OPS victims 'deserve' their fate! This, as we intuitively know, is improper and impolite, if not disgraceful and impertinent.

The political philosopher and sociologist Luc Boltanski claims that such politics of pity employs a specific logic, a so-called topos (1999). This notion is derived from classical rhetoric, which posits a topos for a kind of argumentation with both a cognitive and an affective message. In the case of OPS we might think of the terms 'OPS victim' and 'painful case'. These terms express a cognitive claim – someone has become the victim of a wrong act by someone else – as well as an emotional message: the unfairness and misfortune done to someone (e.g., Skhlar 1990). This relates to the topos of 'accusation'. What can a spectator do when he sees the suffering, but remains at a distance and is unable to do anything about it? He can become 'indignant' or even 'outraged', prompted by pity. Then pity is no longer without power because it takes up the weapon of anger – a deep emotional feeling which strives for action. Because the offender and the victim, or employer and OPS victim, coexist at great distance, verbal violence dominates: the charge of accusation, which is geared to the doer rather than to the 'unfortunate'. However, in such a case causal relationships need to be construed between the doer and the victim. If doer and victim are near to each other, it is easy to explain their causal relationship. Yet it is all the more difficult when the causal links are longer or less obvious, as in the case of OPS. The indignation and accusation, then, requires an antagonistic vocabulary, in which the speaker addresses not only the 'general public' that needs to be persuaded and aroused to action, but also the antagonist, the employer. This leads to the format of the 'dispute', to writings that 'accuse', 'disclose scandals', to pamphlets and brochures that in an agitated tone of voice denounce the injustice and publicly accuse the offender.

The case of OPS shows strong evidence of this logic. The 'pamphleteering style' – found in brochures, newsletters, reports and press releases – underlines the significance attached to the 'OPS case'. Offenders and victims are identified explicitly. In the Netherlands, the FNV Construction Union, as the most implicated division within the FNV federation, led the battle against OPS. In 1991 this division was closely involved in the establishment of the OPS Association (Vereniging OPS), the interest group of OPS patients which is aimed at rendering their status as 'victim' visible. OPS patients would in fact be present at meetings, conferences and news conferences to underscore their sorry fate. But the offenders were rendered visible as well, as evidenced by earlier quotations above. In related cases – involving, for instance, 'burnout' and 'mental fatigue' – one encounters similar antagonistic vocabulary. In the brochure *Beating Burnout*, for instance, one can read: 'Burnout: a disease that occurs evermore often, but is still not taken seriously ... Our members are not left out in the cold! FNV Confederates do take this disease seriously.' (www.bbzfnv.nl, 4/02/2003) Yet there is also ambivalence because in the same brochure, a proposal is made for an approach geared to individual employees ('emotional energy-shots') as well as to individual employers ('hints for leaders'). Conversely, employers and insurers equally deploy a highly antagonistic vocabulary. They responded to the FNV initiative to set up an OPS agency as if stung by a wasp. Employers considered it a 'sick' initiative, while insurers predicted dire circumstances (www. proterra.nl, 24/01/2003). Employers feel that often they are faced with the consequences and therefore they argue for more responsibility among employees and government. In an interview, a staff member of FME-CWM argues that although the cause of many mental complaints 'mainly lies in life's pressures, there is a large chance that the government will opt again for the easiest way and issue new rules on work pressure. Thus a social problem is again one-sidedly blamed on the employers' (Metalektro Profiel, May 2000, pp. 10–11) There is cognitive content here but also an accusing moral tone. The accusation leads to an antagonistic situation. 'Being right' becomes an instrument in the service of 'being proved right'. Hence parties deploy experts to be put in the right.

If it is hard to quarrel about the suffering of individual people who have OPS, quarrelling about who is to blame can be endless. There is uncertainty, but the logic of the juridical situation forces one to identify an offender who can be held responsible for the suffering. Is it the employer? In what measure has the employee been complicit? In what degree did other problems, such as the employee's home situation, play a role? No wonder that juridical procedures take so long.

The protection paradigm and its effects

The protection paradigm searches for juridical solutions to problems involving work and disease: legislation in which standards are defined; and the attribution of responsibility is assigned to employers via, for instance, the reversal of the burden of proof. Conflicts are to be resolved through the courts. But, as Schuyt puts it, the road to legislation leads to 'digital measures in an analogue world' (Schuyt 1982). The law requires straightforward yes/no decisions, whereas the world to which they apply is complex, uncertain and ambiguous.

As a consequence, subtle but relevant distinctions disappear from view. First, in practice it is not always clear who in fact is a victim. Of the many thousands of employees, only some are diagnosed with OPS. Of those not so diagnosed, many can also be marked as 'painful cases'. A substantial number of OPS patients consider themselves victim in the sense that they have a fixation on their illness, in part because they initiated legal procedures or are in conflict with their employer or the social security agency. In this respect, psychotherapy and treatment are geared to let people gain more distance from their ailment and focus on what they can still do and the ways to exploit their possibilities. There are also various 'sorts of victims' among OPS patients: employees who in addition to OPS have a psychiatric or psychological disorder, who are illiterate or feebleminded, suffer from another occupational disease such as back complaints, or who are exposed to noise or have other health problems (Werkgroep Collectieve Compensatie 2003). The legal practice of occupational harm also gives rise to a wholly new type of 'victim': the assertive, active, well-educated employees who, accompanied by lawyers, press charges against their employer and his insurer.

A second distinction that gets lost is that harm or suffering presents itself in quite various ways or is precisely pushed into the background. The domain of work and health in the Netherlands is mainly geared to issues related to absence and reintegration. Individual and collective prevention on the work floor as well as prevention aimed at specific categories of occupational diseases have largely been neglected. As a result, much harm remains invisible. Genetic and other diagnostic tests could be used to target specific disorders, such as among carpenters or welders with an asthmatic predisposition, hairdressers with metal allergy, laboratory-animal caretakers with allergies and employees in the timber processing industry with allergies for dust or wood chips. In such cases, it would be appropriate for employees to be able to have themselves tested, for instance, before they start training. Currently people have the training and then enter the profession, after which the suffering begins.

A third fading distinction is the work versus private life distinction which, though perhaps clear from a legal standpoint, tends to be tangled in actual practice: in fatigue on the job, work-related issues (working in shifts, uncertain job prospects, flexible working options, job conflicts) and home-related issues (raising children, care for ill/disabled relatives, household responsibilities) prove to be major risk factors in work–home conflicts (Jansen 2003). In consideration of many other factors – such as a long commute, work planning, daily care for children and family members – what counts as private or public is precisely subject to negotiation, both in a situational sense (at a social level and within companies) and individually (within relationships, families and extended families). Many employers acknowledge this as well and provide more attention to home situations, for instance, in the context of 'quality entrepreneurship' or 'HRM policy'.

The protection paradigm has no room for discussing such subtle distinctions in normative terms and inasmuch as they are addressed in court cases, only the judges are supposed to evaluate them, not the parties that are pitted against each other. This has a major *social* drawback when it comes to processing experiences involving occupational risks and related changes in scientific knowledge, technological means and the reflection on work and its organization (cf. Chapter 7), society is left merely with the option of legislation and jurisprudence. Thus the protection paradigm makes for an impoverished system, socially as well as in an administrative sense.

The future of genetics and work

In this chapter I argued that the way in which the field of work and health deals with protection against risks, and hence with 'damage' and 'suffering', renders visible the implicit assumptions on the responsibilities of stakeholders. These assumptions are based on a distinction between the individual and social factors of occupational disease: the individual responsibility of employees for their health versus collective protective measures for their health. Insisting on this distinction, the protection paradigm hides from view a practice in the field of work and health in which new relationships are developing between the individual and social factors of occupational diseases. This practice entails new responsibilities for employers, employees and other parties and the emergence of new normative issues. By contrast, the paradigm of protection puts a brake on social learning processes to do with genetics, prevention and the workplace, as well as on the development of, and dealing with, the new responsibilities of the parties involved.

Social experiments in the area of work and health will be necessary because of the growing supply of new genetic tests and techniques for the diagnostics, treatment and prevention of disease. One may object that this invitation to experimenting is utterly lacking in a sense of reality because genetic tests are hardly used in the workplace in Western European countries. The same applies to the United States, as indicated in reports from the 1980s and 1990s (see Chapter 7), even if in that country jobs are often tied to health and life insurance, meaning that employers who are pressured by insurance companies may be more inclined to implement genetic screening so as to determine their employees' risks of disease. There are clear signs, however, that this situation is changing. A survey conducted in 1998 by the American Management Association revealed that 10 per cent of employers routinely test employees for genetic predispositions and that this percentage is rising. A new trend is visible in Europe as well. A study conducted by the Institute of Directors in England in 2000 indicated that 50 per cent of the 353 CEOs questioned approve of genetic screening for occupation-related disorders, at least if the employee voluntarily agrees to it, but 16 per cent feel that it should also be possible to make it mandatory (European Group on Ethics in Science and New Technologies 2003). These shifts in the thinking of parties involved in the field of work and health may well be accelerated in view of the rapid technical advances in genetics. Although today genetics perhaps offers few specific handles for a targeted prevention policy (Van Damme 2000), while genetic research of possible individual sensitivity to solvents provides little hold (Wenker 2001), this situation may change rapidly. In fact, it will increasingly become possible to define and monitor many disorders simultaneously and in a cost-effective manner; our complaint- and disease-based medicine is slowly making way for a medical practice that is geared to genetic risks and the prevention of diseases (see Chapter 1). This 'predictive' medicine will comprehensively apply information on genetic traits, lifestyle aspects and psychosocial factors, in the area of work and health too. This puts more pressure yet again on the relationship between individual and community, and hence between work and private life. The normative shortcomings of the protection paradigm do not just count in the use of genetic testing, but for all forms of predictive testing and prevention geared to medical and psychosocial risks in the workplace.

The protection paradigm hides the above-mentioned shifts from view, thus depriving the parties involved in the field of work and health of potential social learning processes. Genetic and other predictive tests will create new situations in any case, making it only sensible to capitalize on such opportunities for experimenting. Otherwise, the only available alternative for the protection paradigm may establish itself as its rival if not actually replacing it: a practice fully fostered by the premise of individual responsibility. In a way, we have left this paradigm behind already: it belongs to nineteenth-century liberalism/capitalism, and here we need not rehearse the reasons why this paradigm was left behind, to be replaced by the welfare state and its protection paradigm. These reasons are, of course, well-known. The paradigm of individual responsibility does not allow for organizing solidarity. Similarly, the more modern forms of individual responsibility for health are hardly attractive, but are nevertheless making headway both in the clinical context, in which the autonomy principle is put centre-stage, and in the 'health promotion' context, which has embraced the notion of the individual's responsibility and attention to lifestyle. It is but a further step to attributing responsibility to individuals for their health on the job without supplying them the means to be able to do so. An analogy may be helpful here. In many respects people are responsible for their safety at home: they are expected to lock doors, install smoke detectors, put in proper wiring and so on. On airplanes, however, others are in charge of safety measures; airline companies and governments organize all sorts of regulations such as air traffic control and monitoring, without plane passengers having any input. The situation of many employees is more similar to that of a passenger on a plane than that of a homeowner who bears responsibility for a safe home (Mertens, 2003).

However, changed working conditions and changed scientific/technical means *together* may change the position of employees, which may cause their responsibility to shift from that of the passive plane passenger to that of the active automobile driver who has to buckle up and adapt his speed to the traffic situation. This, however, requires new models of thinking and acting in the field of work and health. And it calls for new forms of facilitating social learning rather than merely via legislation and jurisprudence.

What is needed are new normative views. After all, genetic tests function in a practice that exists in part by the grace of normative principles. For instance, Van Damme argues that

> the possible relevance of genetic testing – as of other testing for predicting or preventing disease in the workplace – is not an intrinsic property of these tests but is largely affected by the context of the testing practice: why is it done and who does it are two of the key questions in this respect. In order to answer these questions, key goals and objectives in occupational health practices and the professionalism and the role of occupational health practitioners required to meet these goals and objectives have to be identified.
>
> (Van Damme 2000, p. 24)

As indicated above, genetic and other predictive tests provide opportunities for preventing harm and suffering among employees with asthmatic susceptibility, allergies, skin diseases or other disorders. Whether the opportunities are realistic and should be used depends on many factors. This involves technical and medical questions: how accurate are these tests, how accurate can treatment or prevention be at all and how effective is the care provided? It also involves normative questions: how should suffering, harm and victimization be defined, legitimized and regulated? The future of prevention, genetics and work is determined not only by the available tests and their effectiveness, but also by normative questions about the goals of work and health, the organization of the labour market, health care and our social system, and the distribution of responsibilities and roles in those institutional terrains. The protection paradigm offers no productive frame for discussing these issues. It is not enough merely to mention genetic screening and monitoring as a deterrent, as example of what has to be prevented, or as a boundary that ought not to be crossed.

How the new principles for prevention, genetics and work should look is unclear for the time being. Two general directions can be identified.

First, the jurisdiction involving liability for occupational risks has itself to be informed by innovation, for, as Barendrecht submits:

> the justice system is still organized as a written exchange of points of view, without real dialogue, and with formalistic procedural rules. Those who ever worked with it simply feel that this nineteenth-century system is outdated.
>
> (Barendrecht 2004, p. 25)

Second, the prevention view of occupational risks should be drastically renewed. One could think of experiments with 'prospective responsibility', in which compensation for occupational diseases – arranged either publicly or privately, subsumed under new forms of risk insurance or not, settled at the company or branch level or not – are used to make finding a new job easier, facilitate returning to one's job, and stimulate other reintegration programmes. It should also be possible to experiment with new forms of individual or group-oriented risk prevention. One might think of social experiments with the use of genetic testing for diagnostics, treatment and prevention of specific disorders in which employees can participate at various stages of their professional career, based on, for instance, voluntariness, full information supply and sound agreements on the use of information by employers or other parties such as insurers. As in law so here too, it should become normal in the area of work and health to experiment with institutions, change them drastically, or even close them down.

The question arises whether the invitation to experiment with genetic and other predictive testing in the workplace is accompanied by large social risks, which is not inconceivable. Genetic and other predictive testing on the job, either in the form of determining individual sensitivities to diseases (genetic screening) when selecting, hiring and positioning employees for specific jobs or in the form of medical surveillance and genetic monitoring, *may* lead to social exclusion and discrimination of employees. However, this risk has to be weighed against another risk: that by not experimenting with and ignoring new developments (owing to the protection paradigm), we will be faced with a situation in which the only alternative is a modern form of nineteenth-century liberalism: renewing the old thinking in terms of individual responsibility and the all-out erosion of social solidarity.

9

Learning from the Work that Links Laboratory to Society

Gerard de Vries and Klasien Horstman

The 'unknown' practice of genetic testing

The introduction of genetic testing raises major questions for society, as not only opponents but also proponents of the new techniques acknowledge. Most of these questions are fairly easy to formulate. Which tests are socially acceptable, or even desirable, and which ones have to be rejected? How will detailed knowledge of individual health risks affect the way we organize our lives and make choices in relation to work, diet and lifestyle? What about insurance? How does knowledge of genetic risks affect the way in which society perceives disease and disability? Should we expect a shift in the balance between, on the one hand, individual responsibility for health and, on the other hand, solidarity with those who are ill or disabled or who have a high chance of developing a serious disorder? And if the projected future is not of our liking, what can be done to steer developments in predictive medicine in another direction? Who is in a position to judge with authority on these matters, and who is in a position to take political initiatives if they are needed?

The list of questions is a long one indeed. Answers, however, are not easy to come by. In some cases, this is due to the fact that conflicting normative principles are involved and sometimes because the issues require expert knowledge. But the overwhelming reason that we are in the dark is another and quite straightforward one: the practice of genetic testing is of recent date and there is still very little relevant experience we may rely on to make educated guesses. Although recent decades have seen an enormous increase in the field of genetic knowledge, our understanding of the *practices* in which this knowledge plays a role is still very limited indeed. Inasmuch as we have experience with these practices, it is mainly confined to the field of prenatal diagnostics and research of a few rare, often monogenetic, diseases. The extent to which the available experience can be extrapolated to developments that seem to be in store is hard to assess. Discussions about the future of genetic testing therefore have a highly speculative character.

In Chapter 1 we argued that in the 1980s, the dominant mode of the early public debates on genetic testing tended to be an exchange of hopes and fears. Gradually the debate has shifted to another format. The piecemeal approach that emerged in the 1990s recognizes that the issues around genetic testing are complex and that it does not suffice to speak about them only in global terms. Instead, this approach considers the various tests separately and assesses their reliability and specificity, the conditions under which they might be used, as well as their specific advantages and disadvantages. Based on analyses of the available medical knowledge and of the ethical aspects of specific tests, procedures for responsible professional conduct have been introduced.

The contributors to this volume have chosen another philosophy as their point of departure. With proponents of the piecemeal approach, the authors share the view that speculation-based efforts to assess the social and ethical issues of genetic testing should be approached with a lot of scepticism. Quality public debates seldom thrive on exchanges of science fiction scenarios. But where the piecemeal approach concentrates on the analysis of separate tests and the formulation of rules that have to guarantee responsible application of each test individually, we emphasize that society has to *learn* to deal with these new technologies and that therefore our first task as a society is simply to collect and evaluate *experience* in this field – preferably, broad-based experience.

Our ignorance in the area of the social role of genetic testing is still substantial and there is little doubt that this will continue to be so in the foreseeable future. New developments and technological possibilities are likely to require attention. Moreover, society is evolving, and hence the role genetic tests can play. Chapter 7 has illustrated this by exposing some of the complexities an assessment of the future role of genetic testing in labour contexts has to meet. Not only do we have to discuss the technological aspects of such tests, but the assessment also depends on projections of long-term developments in the way labour is organized. Such future scenarios, however, are highly speculative in nature and they should be approached with sustained scepticism.

In this book, therefore, we have characterized genetic testing in the first place as an *'unknown'* practice. This premise implies, among other things, another perspective on public policy issues from the one advanced in the piecemeal approach. In this chapter, we will further develop this perspective to argue in more detail how the currently dominant piecemeal approach falls short, and also to make suggestions as to how experiences with existing and future practices of genetic testing can be *organized* to stimulate social learning processes.

First, however, we will recapitulate some of the issues addressed in the preceding chapters. This may help to clarify *in what respects* the practices of genetic testing are 'unknown' – not only for society as a whole, but also for the experts active in this field and for the people who have had genetic tests performed. All of us tend to take much of the work involved in genetic testing practices for granted – and hence the uncertainties and responsibilities that come with this work. Relevant features are rendered invisible, not because someone has reason to hide something, but because our common discourse on practices that feature technology narrows our view. First and foremost, we need to make explicit what is implicit in genetic testing. If we are to discuss genetic testing realistically, it is crucial to articulate what is actually going on in practice.

Making invisible work visible

When discussing genetic tests, there is an understandable inclination to focus on the individuals who have them performed and on the physicians in charge of conducting them. The normative problems that arise are subsequently formulated in terms of what the actors we encounter should do, or refrain from. Should a particular test be introduced? If so, how should we regulate the relationships between physicians and those who wish to be tested? How does one guarantee that the patients involved are free either to accept or to reject the test offered?

Framing our problem in these terms, however, comes with the risk of disregarding other questions that deserve our attention. To begin with, the language we use to formulate our concerns is misleading us. The word 'testing' suggests that genetic testing is a matter of standard procedures that quickly establish someone's susceptibility to contracting a specific disease in the course of his life. As became clear in previous chapters, this is hardly ever true. Many genetic tests do not provide certain results, but information on the *likelihood* that someone will develop a disorder in the future. Where tests do provide a high level of certainty, as in the case of DNA diagnostics for monogenetic diseases and chromosome testing, often little can be said by either about when a disorder will reveal itself or the seriousness with which it will do so. In some cases, such as genetic testing for FH, medical experts hold quite differing views on the clinical relevance of a detected mutation (cf. Chapter 4). When in chromosome testing there is an unsought test result, its clinical significance is often unclear (cf. Chapter 2). To establish a definite outcome, much deliberation is needed on the result's practical meaning.

If we do not focus on the situation in which a result is established but consider the trajectory that precedes it, a host of new uncertainties come to light. And here too, they give rise to difficult dilemmas and hence extensive deliberation (cf. Chapters 2 and 3). If chromosome testing may have become a matter of routine by now, in the practice of prenatal testing unexpected and unsought results regularly crop up that call for elaborate weighing of the pros and cons by lab technicians, physicians and the parents involved. DNA diagnostics requires a lengthy process in which one first has to determine whether someone is eligible for such testing, and if so, one subsequently has to obtain cooperation, consent and blood samples from relatives. This requires a lot of work, not just by the physicians who perform the test, but also by those that have the test performed. The question whether a significant result is possible partly depends on whether candidates are willing and able to persuade their family members to cooperate. Similarly, the success of FH detection also relies on the collaboration of family members (cf. Chapter 5). Acting on test results as a way to escape one's biological destiny, such as through long-term medication, subsequently requires serious efforts on the part of the family: parents, for instance, have to teach their children how to live with the disease that runs in the family. This rarely involves a simple task. Those who think they merely have a test performed, then, are fooled to some extent. Being tested is not a matter of just passively waiting for definitive results, but requires a lot of effort, not only from medical professionals but often also on the part of the person to be tested.

Testing or the use of tests is always a matter of work in which many parties are involved: in each case it concerns complex processes of *judging*, of deciding on an array of details and even, in some cases, on major issues of life. Along the way, many questions have to be answered. Those directly involved – physicians and clients – have to face countless choices and they will often have to take steps that they neither planned nor could have anticipated. Is the client who opts for DNA diagnostics also willing to contact family members with whom she had broken years before? How should one react to a family member who does not want to cooperate? How is an attending physician to respond to a client who, during the testing process, changed her mind and, pending the result, announces that she has decided against having an abortion, or who in light of the uncertainties that have arisen makes known that she is about to decide to have one? All the time and effort it takes is rendered invisible when we speak of a 'test' as if it involves a standard procedure that merely has to be performed to obtain a definitive result on someone's predisposition.

It will be clear that this invisible dimension is hardly typical of clinical genetic practices alone. For example, in insurance, which from the outside appears to be a branch of business ruled by strict formalities and economic rationality, similar issues arise (cf. Chapter 6). Companies and underwriters who sell life insurance policies may face unexpected problems from the information provided by clients. Which issues require more information before decisions on acceptance and premium can be taken? If the acceptance process appears to drag on, the company runs the risk of clients deciding to cancel their application and switch to a competitor who does not delve into their medical history as deeply. At any point of the process, the situation that presents itself has to be judged anew.

Organizations such as clinics and insurers of course have specific strategies available for anticipating the problems they face. A clinic's policies have been formulated in inter-collegial meetings by its medical staff; insurance company staff members who decide on acceptance follow rules fixed by the management. Yet time and again exceptional cases or borderline cases show up that call for more deliberation.

Precisely because testing involves a process of judging, its performance requires professional expertise. This is not because experts have learned rules they can follow blindly, but because through training and experience they have gained expertise that enables them to respond to unexpected situations flexibly and judiciously (Dreyfus in press). A physician who has no feel for the contingencies with which he is confronted will fail to do a good job. Apart from professionals, however, genetic testing involves other people – patients, clients, relatives – and they are fully implicated in this practice as well; they too have to judge or grapple with unexpected events. Which roles do they have and how do they obtain the competencies needed for responding adequately? If we aim to probe more deeply and aspire to do more justice to the normative issues involved in genetic testing, we will first need to make the invisible work that is inextricably linked up with its evolving practices visible.

Distributed responsibilities

In this book we did not consider genetic tests as procedures that supply ready-made results. Instead, we geared our attention to the *work* that goes into realizing test results. Our aim is to *make public* what matters in a wide variety of existing practices associated with genetic testing. Numerous questions that have to be dealt with along the way and that require discussion among those involved popped up. With the work rendered visible, the responsibilities the various parties have to take on in the *process* of genetic testing come into view. Professionals who are faced with technical questions in prenatal diagnostics, for instance, already anticipate the decision on whether or not a pregnancy should be terminated – this issue not only arises at the end of the trajectory when the parents who asked for chromosome testing have to decide about an abortion (cf. Chapter 2). Parents who are informed about an FH testing result will sooner or later have to address the problem how to teach their children how to live with the detected disorder (cf. Chapter 5). Speaking about 'testing', then, refers to a *process*, in which decisions have to be made on many fine details. The responsibilities at issue are distributed in meticulous ways. Although we tend to ask questions about 'the' responsibilities at stake, it should by now be clear that these responsibilities are *distributed* in time and among the various actors involved.

Whether the genetic testing process *as a whole* evolves in a responsible way – and hence whether or not this practice should be deemed acceptable in individual cases – depends on whether the work that needs to be done and the responsibilities that *gradually* surface are distributed and articulated in reasonable ways. The normative issues triggered by genetic testing are therefore both simpler and more complicated than the questions that usually arise in public discussions on this topic and that are covered by the piecemeal approach. They are simpler because they apply to details that can be addressed step by step, along the way; they are more complicated because it is difficult to form the global picture on which to found general judgments from those countless details.

In the piecemeal approach the debate on genetic testing is divided into a discussion on the medical merits of a particular test – a topic typically conceived as a matter that primarily has to be decided by medical professionals – and a discussion on the ethical aspects and on legislation and regulation aimed at ensuring the clients' autonomy in making the final decisions on the consequences of a test result. However, once we have become aware of the large number of decisions that have to be made in the course of the genetic testing process, we will no longer think that normative problems concerning genetic testing only surface in the final stage of that process, when a client has to face the question what, given the outcome of the test, has to be done. Normative questions are distributed along the entire trajectory. Actually, autonomy of the patient is barely at stake throughout the practice; by the time final decisions have to be made, doctors and their patients have already discussed the issues and dilemmas involved at length. By exclusively confining the ethical issues to the question of how to guarantee the 'autonomy' of individuals, one neglects the many judgments that have to be made as a specific testing trajectory is unfolding, long before final results are available.

What is true for individuals who are confronted with a technology also applies to society at large. Most people are inclined to speak of technology as a ready-made, as a gift from the scientific world to the public, which they may gratefully accept or perhaps reject or ignore. At this level again, however, the work needed before a technology may begin to deliver useful results is rendered invisible. Regulations, procedures and economic concerns are not side-issues; instead, these various aspects are fully part of the environment in which a technology functions. Essentially, this environment evolves together with a new technology – in processes of mutual influencing, or 'co-evolution'. This dynamic may be illustrated by Illich's example mentioned in Chapter 1. The invention of the automobile did not yet imply the ready availability of motorized traffic of course. Before the notion of motorized traffic made sense, it was necessary to construct suitable roads, build gas stations, pass traffic legislation and organize all those countless other matters that we are inclined to take for granted when we consider an automobile a useful means of transportation. Typically, the acceptance and embedding of a specific technology involves a lengthy process – and the same applies, for that matter, to the rejection of a technology. This also applies to genetic testing. Whether or not 'society' accepts a specific kind of 'test' may trigger heated debate, but here, too, a major part of the work – the issues and judgments at stake – is made invisible by the way in which this particular question is framed. Here, too, intricate relations between technology and society have to be considered.

In the 1980s, public debates were conducted on the question whether introduction of genetic testing would lead to the emergence of a 'genetic underclass', of people discriminated against and excluded from work and insurance on the basis of their genetic profile (cf. Chapters 6, 7 and 8). With this doom scenario looming, some countries – like Belgium – have issued legal bans on the use of genetic information for life insurance. In other countries, for example the Netherlands, insurers have voluntarily curbed the use of genetic tests for insurance. Employment of genetic tests for medical examinations for job hiring is also restricted by stringent legislative regulation. In the context of labour relations, contrary to the spirit of the age, the protection paradigm was adopted and autonomy explicitly rejected. Once we consider the actual functioning of the measures taken, however, complications turn up. The Belgian ban on the use of genetic information has been criticized for lack of clarity: does the ban also apply to the longstanding and common usage of information about the medical history of family members? Moreover, the prohibition may invoke more emphasis on the risks tied to lifestyles – whereby the incorrect notion may creep in that while genetic predisposition in individuals is given, lifestyles are a matter of free and individual choice for which individuals can be fully held liable (cf. Chapter 6). For genetic testing in the selection of job applicants, scrutiny of the details of existing practices is not possible for the time being. Such tests are hardly utilized – perhaps not so much because of the legal ban mentioned, but rather because usable tests for job hiring examinations are scarce (cf. Chapter 7). However, once we consider the problems that emerge in similar situations, such as the complications involving the protection of employees against so-called painter's disease, scores of snags reveal themselves (cf. Chapter 8). Whether someone contracts this disease depends on many factors: his predisposition and the degree to which an employer complies with the legal regulations for protecting his employees, but also the caution with which an employee follows safety guidelines in the workplace and his exposure to hazardous substances outside his job environment, as well as private and psychological factors. The fact that cases brought before the courts by employees who want to hold their employer liable for their health damage may drag on for years suggests how complicated the issues involved are. If attributing responsibilities were a simple matter, courts would be ready to judge in these matters much faster than they do.

How genetic susceptibility and environment factors interact is a tremendously intricate biological and epidemiological problem, and its study is full of theoretical problems and methodological pitfalls. Much is still unknown and one may well have serious doubts whether the issues involved will ever be fully unravelled. If we address how biological fate and lifestyle risks interrelate in the course of an individual's life, the uncertainties only increase. Normative discussions, therefore, are hardly served by pitting the autonomy principle against the protection paradigm. Both are of interest. If individuals choose a specific sort of life, to paraphrase a famous line from Marx, they make their own life but they do not make it as they please; they do not make it under circumstances chosen by themselves, but under circumstances directly encountered, given and transmitted from biological and social conditions. If we want to do more than just reiterate this global, and therefore rather inconsequential claim, we will have to go the tedious way that close attention to details and contexts entails. We will have to address how various parties (medical professionals, patients, insurers, employers, employees, unions and government organizations), in changing configurations, take on responsibilities by formulating judgments and performing actions that are geared either directly or, more typical, indirectly to the global normative questions at stake.

The key normative concern that arises regarding genetic testing, then, is not whether such tests have to be made available and if so, how everyone's freedom to use them or not remains guaranteed. The real question that needs to be answered is how society can be organized so that it learns to deal with practices in which professional expertise plays a central role and in which many normative issues arise that have to be solved by widely divergent parties – normative issues that often are not even recognized as such because they present themselves as technical or practical problems. Moreover, this concern applies to practices in which, in many cases, we do not yet quite know which knowledge, insights and skills are needed and which judgments are sensible, while in these very practices, today's accepted routines may prove to be inadequate tomorrow. To seriously address the normative issues involved in genetic testing, we have first to acknowledge that we are discussing a practice that in many respects is still 'unknown'.

The issue, then, that genetic testing challenges society to address is how it is possible to *learn* to deal with the many uncertainties and responsibilities that emerge in the context of the practice of genetic testing. This is a political and philosophical problem to which the usual answer – regulation and control and pleas for freedom of choice – does not automatically constitute the most fruitful reaction.

Exit, voice and loyalty

Faced with practices in which a large number of judgments play a role and the responsibilities are cluttered, our first reaction often is a call for more rules and supervision, meaning standardization of decision-making and accountability. For fear of excesses and irresponsible decisions, diversity is curtailed.

To be sure, standardization, explicit rules and transparent accountability have major advantages. Bureaucracy offers legal security, protects weaker parties and allows for retort in cases of injustice. No one looks forward to a situation in which being issued a passport depends on the personal judgment – or 'professional insight' – of the county clerk that happens to serve us. In such situation, we prefer the predictability of rules and bureaucracy.

Standardization of decision-making and accountability also has evident drawbacks, however. For one thing, it comes with the risk of a myriad of rules. The call for transparency often leads to an explosion of control measures that are accompanied by new ambiguities, and an imbalance between the time and energy devoted to performing tasks and their monitoring. Moreover, there is the risk of professional expertise being crushed between government-induced control and clients who insist on their rights. This, of course, is not only a concern for the medical world. In nearly all sectors of the public system, professionals – whether it involves teachers, employment experts, or youth welfare workers – have to face up to the situation that their professional responsibility is undermined by a political mania for regulation on the one hand and increasingly assertive citizens on the other. The effect is that the professionals are 'tamed' (Horstman 2000; Tonkens 2003). The confidence in experts as persons who are specialists on the matter is replaced with trust in the rules they follow – their instruments, standards, procedures – and the possibilities of monitoring their proper use.

The call for regulation and control not only leads to a situation in which the job satisfaction of professionals goes down because they feel restricted in their expertise and judgment. There are also losses for society. Professionals derive their social role from the fact that they can tell something that other people do not know. If for fear of uncertainties their job is moulded to exclude each and every surprise, experts are actually made superfluous and the opportunities for society to learn from experiences will be reduced. After all, by standardizing practices and making everything alike, we no longer have to look around because wherever we look, it is all the same anyhow (Horstman 2004). Only where differences exist may we evaluate and eventually, based on experience, make proper choices.

Not every situation is suitable for routines, rules and the rituals that follow from the call for transparency and control. This applies in particular to contexts where innovations occur, uncertainties exist, boundaries are vague or unknown for the time being, or surprises may present themselves. Instead of striving for strict regulation and for restricting plurality, our attention in this sort of situations should go first and foremost to the question of how plurality can be organized and how the competencies needed for gradually taking reasonable decisions are obtained and distributed. Sensible policy-makers will therefore acknowledge the imperfection of our understanding, rather than focus on how to eliminate uncertainties. They will ask themselves how social and institutional *learning processes* can be organized (Van Gunsteren and Van Ruyven 1995).

A theory by the economist and sociologist Albert O. Hirschman, published more than 30 years ago, provides several concepts that can help to specify this question (Hirschman 1970; see also Benschop, Horstman and Vos 2003; Hemerijck 2002). Hirschman observed that while the dysfunctioning of companies and institutions is the rule rather than the exception, most economic theories assume the possibility of a perfect equilibrium between supply and demand. While such theories are based on fairly abstract preconceptions about the rational behaviour of economic subjects in a timeless world, Hirschman argues for an approach that is keyed to concrete social practices – in both the private and public sector – whereby imperfection is accepted as a starting point. This is not to suggest that 'anything goes'. The acceptance of imperfection as something inherent to economic and social life, Hirschman claims, implies attention to the notion that institutions can and must learn. In this context, he asks which mechanisms stimulate learning processes in institutions, so as to improve the quality of products and services. He formulated his answer in terms of three basic concepts: *exit*, *voice* and *loyalty*.

Learning processes are stimulated where institutions receive feedback from citizens and clients. Hirschman distinguishes two kinds of reactions. People can either go elsewhere, switch to the competitor, which is the *exit* option, or give expression to their dissatisfaction, the *voice* option. Both options indicate to an organization that in order to retain its clients and social trust, it has to do a better job.

Voice and exit are usually related to various sorts of institutions. The exit option is mostly seen as the typical reaction by unsatisfied customers of private organizations that function in a free market of competing suppliers. In such a situation those who are unsatisfied can simply go over to a competitor. The voice option is regarded as belonging to public organizations that have a monopoly on specific services. In this context, clients would only express dissatisfaction through complaining, for they cannot go elsewhere.

According to Hirschman, however, it is incorrect to assume that regarding the functioning of private organizations, people only have the exit option. After all, for many private institutional practices it is also true that people do not switch to a competitor immediately when they are dissatisfied about a product or service. Hirschman expresses this behaviour by introducing a third concept: *loyalty*. Even in a free market situation, dissatisfaction does not automatically result in exit. Individuals also enter into a relationship with private organizations for a shorter or longer term, which aside from a functional meaning also has symbolic and emotional significance. They do not break off such ties just like that.

In contrast to prevailing economic theories, Hirschman therefore views loyalty not as a threat to rational economic action, but as a necessary condition for institutional learning processes. If each occurrence of dissatisfaction automatically and instantly results in a switch to the competitor, that is, exit, this would imply a sizable loss of social, economic and organizational capital as well as of learning opportunities. This is why exit, according to Hirschman, is sometimes irrational from an institutional perspective. Even if it can be sensible to *individuals* to reject an institution's services, for *institutions* it is important that at least a substantial number of clients remain loyal.

Loyalty affords credit to an organization and grants it time to respond to failure. Loyalty assumes that people can handle imperfection and live for a time with the problems they are experiencing. However, they ought not to suffer in silence under poor-quality products and services. On the contrary, to stimulate learning processes, people should be able to articulate their discontent. Without the voice option, loyalty constitutes no incentive for reflecting on the quality of services or goods.

Voice is informative: it offers an organization concrete, content-based starting points for improvement. It may take on various forms and range from complaining to exchanging experiences and participation in an organization's design. Where there are only opportunities for complaining, one might speak of 'minimal voice'. Where there is only room to communicate negative experiences, the client is automatically supposed to take the initiative. In this situation, users function as object rather than as the subject of services and they have to put in a lot of effort to be heard. However, voice can also imply the exchange of an array of experiences that do not have to be defined as either negative or positive. Moreover, apart from being an instrument for quality improvement, voice is often an intrinsic part of quality as well. People do not just want to be satisfied; they also want to be taken seriously.

Given the advantages of voice as feedback mechanism in comparison to exit, Hirschman argues for the development of more voice options in private institutions, where thinking in terms of exit is more common and the cost of voice (time, money and effort spent on communication and deliberation) is overestimated while its productiveness is underestimated. He makes sure, though, to point out that we should avoid creating an either/or situation. Learning processes benefit from a specific combination of exit and voice.

Hirschman claims that not only private organizations frequently offer insufficient room to voice, but also public or semi-public institutions. Providing such room would help to improve the quality of their products and services too. Voice matters in the public domain, not only because people – in the absence of an exit option – are thrown back onto it, but also because voice, through the complex character of many public goods, constitutes the pre-eminent means for improving learning processes (Hirschman 1982). After all, in contrast to the quality of soap or refrigerators, the quality of services in institutional domains such as health care and education is hard to determine. This is even more so for newly established services for which ideas on quality have not yet crystallized, clear yardsticks for quality are not yet available, and clients still don't know what they want or may expect from an organization. In this situation, all involved are in fact engaged in a search for the nature of the service and the criteria for its quality. Here, exit should be considered as 'a last resort after voice has failed.' (Hirschman 1970, p. 37)

The emphasis on voice does not imply that we should incessantly communicate our experiences. Effective learning processes, according to Hirschman, benefit from a balanced dose. This applies to exit as well as to voice. If everybody runs away at the first instance of failure, an organization simply does not survive and no learning processes are set in motion. When each and everyone constantly comments about everything, commentary becomes a routine and it loses its significance. Befitting a pragmatist, Hirschman claims that there is no gold standard for the right optimum between silent loyalty, exit and voice. The 'proper dose' has to be established anew each time. Again, also on this score as with others, experience is what matters.

Hirschman's conceptual framework and his emphasis on learning processes suit a perspective on public policy issues that starts with acknowledging that we have to deal with practices which in many respects are 'unknown'. By emphasizing, in addition to the exit option, the importance of voice and its various forms, the significance of organizing a plurality of experiences for social learning is highlighted. Of course, to take advantage of voice and exit, organizations have to be capable of adequately evaluating the experiences with which they are faced to go on to develop products and services that will induce less dissatisfaction among clients. Private organizations that seek to improve customer relations will select new products that meet complaints that have already been articulated. Public organizations that seek to satisfy citizens have to take similar measures. Apart from *organizing voice*, then, improving the *quality of the selection* is of importance. In contrast to thinking that concentrates on the exit option, however, selection is not exclusively something that is left to clients. It is considered as something for which institutions are responsible as well.

The perspective outlined above implies another view on the public concerns about genetic testing. As addressed in the next section, it opens up the possibility to indicate in more detail how the usual ways of articulating the problems raised by genetic testing fall short. This perspective also offers suggestions for a society that has to decide how the issue of genetic testing ought to be tackled. These suggestions will be discussed in the final section.

Where all cards are put on the exit option, few lessons are learned

The advances in genetics confront society with the issue of whether or not specific genetic tests need to be made available. Society will have to select and will want to do so, while it should also enable individuals to make choices about what is offered.

As has been discussed before, the piecemeal approach conceives of this selection process in a specific manner. To start with, it focuses on the medical performance of a genetic test. Are the medical benefits of the test involved clear? Does the test meet generally accepted standards for screening or detection? If the answers are negative, legislation or regulation has to limit or even prohibit the test's introduction. But if the opposite is the case, this subsequently raises the issue of how to guarantee that individuals are able to deal with the test in a responsible way on their own. This again calls for legislation and regulation in which the principle of patient autonomy, rather than normative consensus, is centre-stage. In this way, however, the piecemeal approach articulates the selection process as a problem to which only yes/no answers are possible. Society may or may not reject introduction of a specific form of genetic testing; individuals, when offered a test, may or may not decide to use it. Where the public problems involving genetic testing are thus framed, the discussion is limited to the question of how to organize what Hirschman calls exit options. Moreover, the normative aspects of the selection process are delegated to the private sphere.

Monitoring of the application of adopted tests subsequently occurs primarily along the lines of established professional channels. Public channels for articulating experiences are not explicitly organized. Public discussions, such as contributions in newspapers and publications from patient organizations, potentially contribute to public opinion only *prior* to regulations. Once a test has been introduced, experiences with the test no longer play a role in the piecemeal approach. Discussions informed by experience with the test take place only among professionals or behind the closed doors of consultation rooms. As such, these exchanges have no *public* character. Only when medical professionals systematically start refraining from the use of a specific kind of test, or when clients show up in significantly smaller numbers than originally expected (as happened in the case of the test offered for Huntington's disease), then the question may arise whether the earlier choice was in fact the right one. This question, however, is seldom explicitly posed. In most cases, a test is abandoned only when an alternative with a better performance is available. Of course, if a clinic or some other organization involved in testing unambiguously falls short, there is the option for patients to file a complaint. Ultimately, the medical disciplinary board or the court will then have to assess the failure involved. From a public point of view, however, little is learned from this option: typically, the failure is not treated as a problem of the testing practice, but attributed to the individual care provider, to the organization in which he works, or – in case of complaints about the usage of tests in medical examinations – to an employer or insurer.

By framing the decision-making process in the way the piecemeal approach suggests, society deprives itself of both the possibilities that *voice* may offer and the opportunities for public learning. All cards are put on the exit option. With respect to the medical aspects of the selection process, this is a real option. Professional medical bodies and (in case of population screening) governments have a real choice either to accept or reject a test. Systematic post-hoc evaluation of the tests is however poorly organized. The selection process could be improved by licensing tests explicitly only for a limited period and by requiring systematic post-hoc evaluations before tests are re-approved. With respect to the normative issues, the situation leaves even more to be desired. The exclusive emphasis on autonomy neglects the fact that once individuals have entered a testing practice, they gradually take on responsibilities for what is going on. In their discussions with the medical staff, there will be ample opportunity to articulate concerns. However, because these discussions take place in the privacy of the consultation room, voice is not organized in a way that stimulates collective, public learning. Individual caretakers will no doubt benefit from the comments of their patients, and future patients may benefit from the lessons they have learned. But the process of selection for future tests is not systematically improved. The potential of voice to improve practices is clearly under-used.

Collective public learning through organizing *voice*

Genetic testing is just one example of a range of subjects surfacing in modern knowledge societies that involve new entities, technologies or knowledge-intensive practices that come with unknown features. Even experts still have little experience with these 'unidentified political objects' (Dijstelbloem, Schuyt and De Vries 2004). Confronted with the question of how to respond, the public at large is left empty-handed. As a consequence, political discussions may soon boil down to a hardly productive exchange on the blessings of innovation advanced by enthusiastic proponents and the pessimistic views of opponents that are probably as exaggerated. One way of moving beyond this gridlock is to opt for a road that allows society to initiate processes of collective, public learning. To make genuine selections, first and foremost broad and varied experience is needed.

The proper way to generate and accumulate experience in situations that are in relevant aspects 'unknown' with the aim of collective learning is to consciously *experiment*. The scientific connotations of the notion of 'experiment' should be taken quite seriously here. To experiment, then, involves, first, explicit acknowledgment that there is in fact an experiment, and, second, careful performance and clear communication of what was done, including attention on possible unexpected events. In scientific journals and conferences, the sciences have quite specific institutions for this purpose. To contribute to science, a researcher cannot limit himself to reporting only results; the path that led to those results and the judgments made along the way also have to be communicated; upon request, more details have to be provided and criticisms addressed. Experimenting, moreover, requires variation. When everyone performs the same experiment, little will be learned collectively. Although major experiments are reproduced for checking particular claims, the marginal returns of repeated experiments rapidly diminish.

While doing experiments is an established practice for collective learning in science, institutionalized practices that meet similar criteria are scarce at the public policy level. Certainly, in the 1990s new ideals of governance increasingly required public organizations to be transparent. Generally, however, this is not the same as the openness that is asked from experimental researchers in the sciences. 'Transparency' requirements focus on prior publishing of clear objectives, and on pre-established criteria for assessing results and attribution of responsibility in case of failure. This type of transparency aims to avoid surprises by deciding how accountability will be settled in advance. But experiments aim to do something else. They are set up not to prevent surprises, but to collectively learn from them.

If we are to develop a social equivalent for the way in which specific learning processes are organized in science, the method of establishing 'best practices' offers a good starting point. In this method, existing practices are evaluated for their performance, to help make informed judgments about how best to do a particular job. Although the point of evaluating is to identify the approach that gives the best results so that others may follow the example, it is essential for this method that as regards its input, it requires variation in practices. As such the method of selecting for best practice is at odds with premature standardization and general rules. By making explicit which steps were taken, what kind of work had to be put in, which judgments played a role and how one dealt with contingent events that occurred, the selection process could be improved. To identify 'best practices', we need to assess their specificities. If we want to learn collectively from successful practices, the trajectory along which their accomplishments were realized is at least as relevant as their performances.

In genetic diagnostics the road to results is long and it requires input from many persons – not only from medical professionals but also from people who thought they were offered diagnostics only to discover – probably much to their surprise – that they were put to work. Those who engage in sensible experimenting, then, will explicitly allow that besides professional considerations other perspectives are taken into account as well, and, that in addition to 'hard', objectified knowledge there is 'soft' knowledge that emanates from acknowledgement and compassion, from passionate reaction to unusual events and situations. In such situations, *voice* can be uncertain, cautious, tentative and hesitant. In collecting experiences, it is not only instant opinions, articulate complaints and firm opinions that should count. If we are interested in the experiences of those who are involved as clients in genetic diagnostic practices, we can most of the time leave questionnaires, psychological tests and polling techniques at home. Bureaucraticized information will seldom capture the relevant experience. One does not uncover the surprises and uncertainties that individuals involved in the practice of genetic testing encounter by tabulating their answers to standardized questions. A more productive method, as we tried to demonstrate in some of the previous chapters, relies on extended conversation, long-term observations and in-depth interviews.

Where the goal is to learn collectively, experience has to be *organized* and *channelled*. This is a task that various parties have to take on. Governments may consider it their task to lead the way, and in fact many have done so already. In several Western countries, they have put in efforts by organizing consensus-building conferences and broad public discussions about new technologies. But so far, these initiatives have had little success. They often start from an incorrect premise. By gathering experts and interested parties, one expects to accumulate the knowledge to achieve consensus about sensible policy-making. By juxtaposing existing and already articulated views, a solid base for consensus and regulation is expected to be within reach. To a large extent, this is an illusion. The public policy problem that governments have to face is not lack of consensus, but the fact that the practices involved are 'unknown' and that for this reason, the first objective should be to articulate experiences so that it becomes clearer *about what* exactly consensus is sought. Discussions that aim at formulating common points of view are premature. In the case of practices that are 'unknown', our first task is to draw attention to variation, unforeseen aspects and new facts. We should view public discussion in society with the same attitude with which we appreciate open, critical discussion in science. The issue is not so much who says what or the level of consensus achieved, but which novel aspects of an issue are put forward (De Vries 2003). Rather than aiming for consensus, the objective should first be to widen and deepen various perspectives, with the aim of enriching the quality of the debate.

Apart from government, private organizations can play a major role as well. In the last two decades, patient organizations have increasingly been involved in medical-ethical issues. They, too, can play a major part in the public presentation of a particular cause (Nelis, De Vries and Hagendijk 2004; Nelis, De Vries and Hagendijk 2007). Likewise, unions, employers' organizations and professional organizations of insurers may contribute to public discussions on genetic testing, and some of them already do. In addition, there is a task for medical professional associations. Where there is public debate on new, often technical and complex topics, the input of experts cannot be ignored.

Experiences with new technologies distinctly belong to the public domain. Such experiences should not remain confined to consultation room talk, discussions among professionals involved, or the academic press. They deserve to be made public and this requires attention to the way in which such experiences are articulated and processed. In this area, there is still much room for improvement. The model in which physicians conceive of their public task as merely explaining complex issues to a broad audience is by now outdated. The image of a doctor sitting behind his desk in his white coat to explain medical issues to a TV audience definitely belongs to the past. There is an urgent need for experimenting with new forms. In today's media, there is a substantial risk of the role of the professional becoming marginalized and that professional experiences will consequently remain outside the scope of the public domain. In today's media-culture, laypersons – patients, parents and relatives – are actively invited to articulate their experiences and air their opinions, while experts are at best confined to the role of commentator. Marginalization of expertise, however, should be a source of political concern. This is not because experts know it all. Just like all of us, they fumble in the dark on the social dimensions associated to their field. But experts have experience with practices that cannot be discarded in collective, public learning processes. They can deliver a unique and indispensable input, provided that they realize they are part of – rather than above – society. Their contribution is valuable for the content they – and nobody else – can offer, not because it is delivered from a superior position (Horstman 2002).

There is still much to be learned about genetic testing, both scientifically and politically. The scientific side of that learning process is well organized, but the political, normative, dimension is much less so. Public learning processes, however, require another institutional or administrative model, not the usual one. The question of how one has to *select* and which rules and laws are needed to offer protection on the one hand, and enable autonomous individual choices on supply on the other, is simply too general for the issues at stake in genetic testing. Too little is yet known about the practice of genetic testing to make responsible policies. It is in many respects still unclear what particular shifts the new methods of diagnostics will bring about in the interrelation of the risks of biological destiny and lifestyles and how the various responsibilities implied will be distributed. We still have to invent what constitutes a sensible way of giving shape to the relationship between individual responsibility and solidarity in various sectors of social life. There is an urgent need to rethink the practice of patient autonomy in predictive medicine. The current notion that in the end, patients have an exit option is little more than an illusion in the face of the long trajectory in which they become entangled. Of course, once in a while, patients will decide to terminate testing in the final stages of the testing process, or to take on the consequences of test results (such as an abortion). They should have the right to do so. But given the often immense problems that they have to cope with, new forms of exit should be actively developed. The reassurances in the folder that patients receive when they enter the long trajectory of genetic testing that 'at any stage you can decide to opt out and you will have the final say when it comes to deciding on further action' is simply not enough. New combinations of *loyalty*, *exit* and *voice* have to be developed for the sectors that involve genetic testing, such as health care, work and insurance. Again, the way to go is to prudently experiment.

Today's dominant style of reasoning on the social consequences of genetic testing, the piecemeal approach, exclusively emphasizes the development of exit options. By contrast, we argue for a systematic strengthening of voice and for new combinations of exit, voice and loyalty. This is primarily a public policy task that will entail experiments with institutional innovation. In this context, the social sciences may have a role to play, in particular with regard to helping to articulate voice. 'The human meaning of public issues must be revealed by relating them to personal troubles – and to the problems of the individual life', C. Wright Mills wrote, subsequently linking this insight to the specific task of the social sciences:

> The problems of social science, when adequately formulated, must include both troubles and issues, both biography and history, and the range of their intricate relations. Within that range the life of the individual and the making of societies occur; and within that range the sociological imagination has its chance to make a difference in the quality of human life in our time.
>
> (Mills 1959, p. 226)

In this book a modest attempt was made to fulfil this task by articulating a variety of experiences in already existing practices of genetic testing. Thus the authors have tried to practise what they preach.

Notes

2 Constructing results in prenatal diagnosis: beyond technological testing and moral decision-making

1 In prenatal diagnosis most of the professionals are women. To prevent any form of identification, *all* observed professionals are consistently referred to in the female.

2 This case study has been composed from various case studies observed in this research project.

3 In relevant literature, grey results are often categorized under the heading 'unexpected findings'. However, this term is somewhat confusing as the common feature of these types of results is not so much the unexpected element, but rather the element of ambiguity (Van Zwieten, Willems et al. 2005).

4 As the large majority (some 95 per cent) of the results in the daily practice of prenatal diagnosis concerns normal outcomes, I applied a specific procedure that allowed me to be present at relevant moments. This way I could do field research without 'hanging round' all the time in the department. I was supposed to be informed in case of any result that *might* be aberrant, that is, with 'grey' as well as 'black' results. All technicians were instructed to call me the moment they observed something in the laboratory that might produce an aberrant outcome. When a technician wanted to consult the cytogeneticist, all parties also agreed not to take any action before informing me. As soon as I received a phone call from a technician, I went over to the department and accompanied the technician to her meeting with the cytogeneticist. As of that moment, I monitored the result in question and, if possible, attended all interprofessional consultations with respect to the outcome: between the technician and cytogeneticist; the cytogeneticist and clinical geneticist; and/or between the clinical geneticist and gynaecologist. I was always present at the weekly interdisciplinary meeting of the entire professional team involved in prenatal diagnosis in the AMC. Sometimes I attended the weekly technicians' meeting.

5 During the observations and immediately afterwards, I made notes of what I heard and saw during the professional consultations. If possible, I relied on verbatim quotes. These notes were subsequently developed into a comprehensive and detailed report of all observations, an observation protocol: i.e., a description of all experiences that were suitable for (external) third parties. This protocol contained factual and direct descriptions of the observed situation; for example, in the form of quotes or indirect speech complemented by my impressions and interpretations. In agreement with proper methodological standards, the different types of entries were clearly distinguished.

6 In order to analyse the observation data, I first structured all survey material per result trajectory. I put all fragments from the observation protocol relating to the same trajectory in chronological order. To this end, I made use of Kwalitan, a software program specially developed for analysing qualitative research data. I reconstructed all observed result trajectories, partly based on

191

additional file surveys. I had access to the relevant patient files with laboratory forms, the result letter and all other correspondence with medical specialists, as well as the minutes of the weekly interdisciplinary meeting. The next step involved writing up the reconstructed trajectories into a comprehensive report, always taking the primary survey data from the observation protocol into account. I submitted the descriptions of the observed result trajectories to a cytogeneticist and clinical geneticist of the AMC Department to check the reports for factual irregularities. I also wanted to know to what extent they recognized themselves in their capacity of health professionals in my account of the observed setting – a form of validation known as *member check*; an important instrument in qualitative research to guarantee the validity of scientific research.

7   The chances that a child with a (familial) balanced structural aberration will later give birth to a child with an unbalanced structural aberration are larger, however.

8   ICSI (Intra-Cytoplasmic Sperm Injection) is a form of IVF in which a sperm cell is injected directly into the egg cell. ICSI is an accepted indication to perform prenatal diagnosis because of a possible increased risk of chromosomal abnormalities.

9   Nuchal translucency measuring is a form of prenatal screening in which the thickness of the nuchal translucency is measured through an ultrasound scan.

10 When chorionic villi sampling or amniocentesis is performed because of an increased risk for a specific DNA abnormality, it is standard procedure – after consulting the patient – to screen the chromosomes as well.

11 The combination of mosaicism of cells with only a single X-chromosome and normal female cells gives a different situation than such mosaicism in combination with normal male cells. In the case of a combination with normal female cells (45,X/46,XX) the phenotypical outcome is comparable to Turner's syndrome, but less severe; girls with this chromosome pattern may have a normal appearance. Mosaicism in combination with normal male cells (45,X/46,XY) leads to a more complex phenotypical situation because there may be ambiguity about the gender. In this trajectory the ultrasound scan was used to determine whether the foetus had normal male genital organs. If so, it would be improbable that this was a case of mosaicism 45,X/46,XY.

12 The triple test is a measurement of three substances in the blood of the pregnant woman, which enables a prognosis of the risk of Down's syndrome.

13 In conventional chromosome testing, as described in Box 2, only cells that are presently in metaphase can be screened, as the chromosomes only become visible during that phase. Interphase Fluorescent In Situ Hybridization (i-FISH) also enables screening of all cells in interphase. This makes possible the simple screening of a larger number of cells in order to answer a specific question.

3   Genetic diagnostics for hereditary breast cancer: displacement of uncertainty and responsibility

1   One example is the series of portraits of individuals from families with hereditary breast/ovarian cancer that in February and March 2003 appeared weekly in *de Volkskrant*, a leading Dutch national newspaper. The Dutch TV

programmes *Vinger aan de pols* and *In de schaduw van het nieuws* also paid attention to hereditary breast/ovarian cancer, on 14 September 2002 and 15 July 2003, respectively.

2   For decades the professional ethics of geneticists has been strongly coloured by the thinking in terms of autonomy. This has to do in part with eugenic practices that in the nineteenth century and the first half of the twentieth century existed in many Western countries, whereby the reproductive freedom of individuals was curbed in the interests of public health. After World War II and Nazi eugenic politics, eugenics was broadly rejected as immoral. By stressing the individual client's freedom of choice, genetics managed to distance itself from such eugenic practices. In 1947 the term 'genetic counselling' was even introduced with the explicit intention of keeping eugenic associations at bay (Bolt 1997, p. 7). Aside from this urge to distinguish itself from past practices, the fact that after the introduction of genetic diagnostics (but before the era of DNA diagnostics) the only available *therapy* was abortion played a role as well. On this issue, too, geneticists were able to by-pass ethical sensitivities by foregrounding the autonomy of those who asked for advice. To realize the autonomy of the client (or those in need of advice), the genetic counsellor would have to take a neutral and nondirective stance; meaning that as a rule, he/she, unlike other medical professionals, would have to refrain from providing explicit advice to the client (Bosk 1993; Fine 1993). For this reason, there has been a discussion for some time in the Netherlands on whether the term 'heredity advising' was not a too directive translation of 'genetic counselling'(Bolt 1997, p. 8).

3   A recent example of an ethical approach to DNA diagnostics as a ready-made 'test' that mainly addresses choices *before* and *after* the test can be found in Gordijn (2004). For example, are parents allowed to test their unborn child to see if it is carrier of a BRCA mutation? And to what extent is it permitted to remain silent about the information derived from genetic tests vis-à-vis relatives? The availability of a ready-made genetic test is simply taken for granted. By contrast, my contribution argues that ethical questions should also address the *technology itself*, rather than ignore it.

4   This chapter is based on fieldwork I did in 2002 and 2003 in an outpatient clinic for Clinical Genetics at a Dutch academic hospital. All women making an appointment with this clinic in the period between February and June 2002 with a question about hereditary breast/ovarian cancer, and who themselves at that point had not yet had breast or ovarian cancer, received a letter explaining the study and requesting their collaboration. A staff member from the clinic would call these individuals a few days after they received the letter to find out if they wanted to cooperate; if so, their name and phone number were given to me and I made an appointment with the person involved. Thus I came into contact with a total of 11 people from ten families: nine women and two men. The two men were brothers and had a shared consultation with the clinical geneticist; the nine women were not related to each other. I interviewed all participants (mostly at their homes) before they had their first appointment in the clinic. The interviews mainly addressed their reasons and concerns for requesting such counselling, as well as their expectations and possible plans. Next, I observed and taped the first counselling interview between these individuals and the clinical geneticist. I was also

present at their follow-up consultations in the clinic. (On one occasion I was unable to attend a consultation and this one was taped; on another occasion I was not informed about a follow-up appointment.) In sum the 11 participants had six follow-up interviews with the clinical geneticist, sometimes together with relatives. Some nine to 12 months after the first interview, I held a second interview with all female participants. (A second interview with both men was cancelled because my research topics proved to play only a marginal role in their case.) At that point seven women had finished their genetic counselling trajectory in the clinic, while two others were still engaged in it. These interviews thus offered a single-moment picture of the counselling trajectory's progress and thereby I mainly foregrounded the relationship between the experiences of participants up to that moment and their initial expectations. Moreover, during the first observation period I also interviewed the two clinical geneticists involved. There are full transcriptions of all interviews and observed counselling interviews.

All transcriptions were analysed for recurring themes. Because of the limited reach of the fieldwork conducted for this study, the findings and conclusions are not representative for all counselling trajectories around DNA diagnostics for breast/ovarian cancer. Nevertheless, the collected materials offer sufficient starting points for exploring the field and questioning specific elements. The themes that might be interesting for further analysis were identified based on comparison: of interviews and observations, empirical material and literature, and empirical material and images of this practice as appearing in public information and public media. By zooming in on these themes, it was possible to reconstruct the practice in a way that tips the usual perspective and introduces productive new vantage points.

5   The notion that a diagnostic 'test' is no isolated technical artefact but implies a social practice has meanwhile become widely accepted in science and technology studies. Latour (1987) coined the term 'black box' for technologies or scientific facts that are seen as given. Much science and technology studies are geared to opening that black box and revealing how much effort is needed to construct a 'reliable technology' or a 'hard fact'. For examples of this in the area of medical diagnostics, see Horstman (1997); Mol (2000).

6   In the Netherlands an estimated 10,000 women contract breast cancer each year. According to the percentages mentioned, some 500 to 800 cases would involve hereditary breast cancer (Voogd, Rutgers and van Leeuwen 2002).

7   Jorde and colleagues report that at BRCA1, more than 200 different mutations are identified and at BRCA2, more than 100 different ones (Jorde, Carey et al. 1999, p. 236).

8   The myriad of potentially pathogenic mutations in a gene (also referred to as 'allelic heterogeneity') occurs in both monogenetic and multifactorial disorders. In many cases this reduces the clinical usability of DNA diagnostics; after all, as long as not all mutations are known, the sensitivity of the diagnostics (the percentage mutation carriers that is actually diagnosed as such) is limited and hence so is its validity. But also, when all mutations can be identified, their clinical significance is not always clear. Burke rightly claims that poor validity does not have to be only a matter of limited knowledge; the large number of mutations may also render a test that detects all known mutations particularly expensive, warranting the adoption of a middle course between

cost and sensitivity (Burke 2002). In the United States this has long been the case, for instance, for CF diagnostics.

9   This way of working is largely based on the protocol developed in the late 1980s for the first disorder for which predictive DNA diagnostics became possible: Huntington's disease. This is a degenerative disorder that usually reveals itself between the ages of 20 and 50, and from which patients will die after 10–20 years. The mutations that cause Huntington are almost fully penetrant, meaning that mutation carriers will almost certainly contract the disease. Initially this contributed to the fear that it would be too much of a burden for people to know they were a carrier, and a protocol was formulated in which much attention was paid to the quality of the client's decision process and the conditions that specialists had to create for it, for instance, in the form of elaborate counselling and time for reflection (Nelis 1998).

10 The quotations in this chapter are as far as possible taken directly from the transcriptions. For the sake of readability, the interviewees' hesitations and the interviewer's brief interjections are mostly left out, while the oral language used is presented here in a somewhat polished version. When words or parts of sentences are left out, it is indicated by an ellipsis; when something is added for clarity, it is enclosed in square brackets. All names used are fictitious.

11 In part at the insistence of the professional organizations of geneticists, it is now under consideration whether or not the term should be extended.

12 Technically, in principle at least, it is possible to test DNA from healthy people directly for mutations. The DNA testing for breast/ovarian cancer that is performed in Dutch centres implies that both BRCA genes are entirely analysed for deviations. Whether they are pathogenic can be assessed on the basis of comparison with already known aberrations in other families (comment via email from clinical geneticist Arends, 3 June 2004).

13 This role for ill relatives in testing healthy relatives is necessary for all diseases where the existing DNA diagnostics has low sensitivity (Burke 2002).

14 My second interview with each of the participants allowed me to identify how they were doing and what decisions they took. Briefly put, they fared as follows:

- In the family of three participants a genetic mutation had been established earlier; the three were subsequently tested and one proved to be a carrier but not the other two (both men); the mutation carrier opted for periodic screening of breasts and ovaries.
- Three participants decided to have no DNA diagnostics (two because they want screening but no clarity about their risk, one because the family is unwilling to participate); all three opted for periodic screening; one of them, however, has not yet picked up the referral letter that has been waiting for her in the family doctor's office for several months.
- Two participants have not yet taken a decision (in one case because a family member who would have to cooperate had meanwhile contracted breast cancer again, while the other is waiting for the clinical geneticist's reaction to the sent-in family data); they have not yet taken preventive measures.
- One participant, whose sister had already tested negative, tested negative as well; she opted for screening of breasts and ovaries.

- One participant was not tested but her (afflicted) mother was; her result was negative, after which further DNA testing of the daughter was refrained from; both mother and daughter subsequently opted for periodic breast screening; the mother also receives monitoring of her ovaries.
- One participant was tested as the only family member because there were no afflicted family members left; her result was negative; for the time being she opted for screening of breasts and ovaries.

15 The 'shared decision-making' model suggests that physicians do not restrict themselves to explaining the information on diagnosis and the available options, after which patients decide, but that both physician and patient exchange views on what would be the best thing to do for the patient – each from their own vantage point – after which, ideally, they achieve a consensus on the steps to be taken (Elwyn, Gray and Clarke 2000). Similar proposals can be found in Bolt (1997), who feels that counsellor and client should be striving for 'inclusive rationality', and in Parker (2001), who emphasizes the importance of 'interpersonal deliberation' in counselling.

16 For a similar interpretation of the process of genetic counselling, see Armstrong and colleagues (Armstrong, Michie et al. 1998). Armstrong and colleagues, however, only considered the counselling *process*, not its results.

17 Clients may, of course, put forward this argument themselves. If they manage to find out that they are a mutation carrier, at least they know that monitoring makes sense. In the case of a negative DNA result, in other words, women no longer have to undergo the unpleasant intervention of regular monitoring. The problem is, however, that this only applies to a small group: namely only those women in whose family a mutation was detected of which they themselves proved not to be a carrier.

18 It would be interesting to make a cost-effectiveness study in which the current situation – in which all women who, based on their family-tree, seem to be at risk for hereditary breast/ovarian cancer are eligible for periodic monitoring (even if they decide not to undergo DNA diagnostics) – is compared to a situation in which DNA diagnostics would be mandatory for everyone wanting periodic monitoring or preventive surgery. That the second situation would be more cost-effective seems doubtful to me, at least as long as DNA diagnostics has such poor reliability. Recent research by Breheny and colleagues suggests that predictive DNA diagnostics may be cost-effective compared to population surveillance, but their investigation pertains to testing of first-degree relatives of *known* BRCA1 and BRCA2 mutation carriers only (Breheny, Geelhoed et al. 2005).

19 For such a scenario, see STG (2000).

20 The issue of the exact nature of 'choosing' is discussed in general terms by Mol (1997). For a developed pragmatic take on ethical reflection, see Keulartz (Keulartz, Korthals et al. 2002).

4   Lifestyle, genes and cholesterol: new struggles about responsibility and solidarity

1 For a historical-sociological analysis of the concept 'cardiovascular disease', see Bartley (1985).

2   According to Hoeg, this was an unintended effect of this study. They precisely were after a causal explanation, finding a 'cause' for cardiovascular diseases, but in 1961 they had to conclude that 'no single essential factor has been identified' (Hoeg 1997).

3   At the time of the start of the Framingham study and in the 1950s and 1960s, personality psychologists also contributed their mite to the explanation of cardiovascular disease. They introduced two personality types – type A and type B – to explain why some individuals with a specific behavioural pattern were prone to cardiovascular diseases. This holistic approach, however, did not become as popular as the risk-factor approach (Riska 2000).

4   If the 1953 flooding of the Netherlands was followed by a comprehensive 'Delta plan' to prevent such disaster from ever happening again, in the 1970s there was discussion in NTvG about a 'Delta plan' for the fight against cardiovascular diseases – a notion that suggests we are all victimized by the Western lifestyle in equal measure.

5   During that process the content of that claim changes and the facts that are eventually produced may substantially differ from the claims that were formulated initially. Conditions are added and refinements and specifications are introduced, which often cause the original claim to become unrecognizable. This hardly alters the idea, however, that claims anticipate their own factuality in order to be able to become facts.

6   The policy plan of 1995, the annual reports of the StOEH from 1995 to 1998 and a WHO report were part of the application of a permit in the context of the WHO: a permit for screening familial hypercholesterolaemia in people over 16. This application, though, was not attended to because the detection of FH was not seen as requiring a permit. In these documents many of the same arguments recur time and again. In my analysis I often use the 1995 policy plan as illustration, but it is easy to find similar texts in the other documents.

7   Reports by the WHO also describe FH as 'a classical monogenetic disorder' (WHO 1998, 2).

8   This is, according to the StOEH, partly an effect of ignorance among physicians. 'The familiarity with the clinical picture and the hereditary pattern of classical FH is unfortunately limited among GPs, cardiologist and internists' (Kastelein and Defesche 1995, p. 7).

5   Detecting Familial Hypercholesterolemia: escaping the family history?

1   A study by Sobel and Brookes Cowan into the effects of genetic diagnostics for Huntington's disease on family relationships has shown that this diagnostic trajectory, regardless of the results, affects the family for various reasons. Those involved, for instance, experienced both the secrecy and the revelation of a mysterious family disease as problematic. Or because the perspective on a short or long life of other family members was adjusted after the results, creating new attitudes and relationships: for instance, one person was criticized more for specific conduct after testing negatively (meaning no Huntington's), while earlier it was tolerated because family members expected

not to live lives of a normal length together. Sometimes anger and reproach came up because someone did not share a specific test result. Sobel and Brookes Cowan speak of 'unfinished businesses' in the family (2000) See also: *American Journal of Medical Genetics*, special issue *Genetic Testing and the Family*, 119C:1 (2003).

2 This is also true of the women in Chapter 3 in this volume.

3 This argument is also used in the context of insurance. If a life insurance company wants to exclude people with FH because of high mortality risk, it is countered that people with a positive FH diagnosis precisely present a normal risk because after the result they use medication that reduces their risk. In a flyer from the Dutch Association of Insurers we can read about FH: 'Whoever would like to purchase life insurance or disability insurance may come into contact with the insurer's medical adviser. This person evaluates the application and thus the risk for the insurer. Accordingly, individuals with (the chance of) FH may have to deal with the medical advisor if they want to take out insurance. The health condition of those with FH is assessed in the same way as that of those without FH. The cholesterol level and the other potential risk factors, such as weight, blood pressure and smoking, determine the risk. It is possible that treatment leads to improved readings. These, of course, are also weighed in the risk assessment.'

4 The names of these families are fictionalized. The quotations are followed by a family name and capital letter (A, B, C, etc.), which refers to a specific member of that family. This allows the reader to trace throughout the text which specific family member is quoted.

5 Chapter 3 by Boenink in this volume reveals how much work women have to do in the family to make genetic diagnostics of breast cancer possible. The question of the role played by families in the diagnostic trajectory of FH has not been studied so far. Compared to the diagnostics for breast cancer, genetic diagnostics for FH has a more routine character: in most cases the result is known after four weeks. Yet it is true that the clinical relevance of all mutations is far from clear (Graham, McClean et al. 1999; Marks, Thorogood et al. 2003). There are milder and more severe variants and the latter may prove to be not so severe after all, while mutations may prove to be polymorphisms that don't do anything. Just like in hereditary breast cancer, there are compensating genes that make sure that individuals with a mutation still develop no high cholesterol or cardiovascular diseases. Conversely, there are people with a clinical diagnosis of FH and a family history that points in the direction of a genetic disorder in whom no mutation is found. In other words, the relation between genotype and phenotype is complex and the participation of families in research is a crucial condition for gaining more insight into this issue and improving diagnostics in the long run. Regardless of the complexity of the relation between genotype and phenotype, the therapy implications tied to a positive result seem highly standardized (as is true with hereditary breast cancer): generally, in the case of a positive result, medication and lifestyle change are advised. This also applies when a positive genetic test result is not reflected in a raised cholesterol level. Medication is unnecessary only when no other risk factors for cardiovascular diseases are present. Against the background of the complexity of the relation between genotype and phenotype, Wiegman and colleagues argue that in decisions on medication for young children who have a mutation, the family history and the seriousness

of cardiovascular diseases in parents should also be taken into consideration (Wiegman, Rodenburg et al. 2003).

6   See the letter by the Dutch Health Minister to the Chairman of the Health Care Insurance Board (College voor Zorgverzekeringen), 19 June 2001: 'Although the assessors [AMC-researchers] express reservations about the high cost per year of life gained that are tied, according to them, to FH screening and the treatment of detected gene carriers, my position is that people with hereditary hyperlipidaemia, given their risk, are eligible for cholesterol-lowering treatment. Given the fact that people with FH can be efficiently detected and effectively treated with medication and lifestyle advice, I support the introduction of a nationwide screening programme in the short run, at least if this is done efficiently, an adequate follow-up of the detected high-risk carriers can be guaranteed. The detection of people with FH can be done through family testing based on the StOEH model' (Borst and Hoogervorst, 2001).

7   A positive test result can give rise to adjusting the dosage.

8   In a study of the risk experience of people who participated in a FH detection programme, it also became clear that various people construct a break in the family history to downplay their own risk of cardiovascular diseases (Senior, Smith et al. 2002).

9   For a long time these debates centred on the question whether the application of genetic tests in children was ethically responsible. Given the dominance of the autonomy principle in the ethics of genetic diagnostics, this is hardly surprising. From that view it is not legitimate that parents decide for their children and it is the task of geneticists to protect children against their parents. Meanwhile, it has become clear that the autonomy principle and genetic testing do not go together well in actual practice because of the inherent family nature of the issues involved. Although the difference between norm and practice is sometimes grounded in the need to emphasize the norm, in this case revision of the norm is deemed necessary. Recently experts have argued for the attribution of more meaning to the family, in a conceptual-normative sense, when genetic testing of children is involved (McConkie-Rosell and Spiridigliozzi 2004). In this article we do not address the normative issue of whether testing children is ethically right. Our central concern is with the meaning of genetic diagnostics in everyday practice.

10   In a study of medication in children, De Jongh concludes that statins are an effective drug in children from the age of 8 and that they do not negatively influence the growth of children in puberty (De Jongh, Kerckhoffs et al. 2002). Others point to the fact that little is known still about the effects of medication (Marks, Thorogood et al. 2003). At the time of this study, in the Netherlands statins were not yet administered as regular medication to children under 18. The children of our interviewees who were prescribed statins probably participated in an experimental study. Bakker and colleagues had already argued several years ago for allowing medication in children (Bakker, Wiegman et al. 1997) and there are signs that the age limit will be lowered in the short term.

11   Given the possibility of making the actual damage in the arteries visible with the help of visualization techniques, the nature of the predictive value of this diagnostic technique and the DNA technology – as well as how these techniques relate to each other in the future – will remain an issue.

12   Tonstad concludes on the basis of a study of 154 parents or parental couples and their children that 20 per cent of the parents report family conflicts and 8 per cent negative changes in the child after being diagnosed (Tonstad 1996). No parents speak of psychosocial problems and Tonstad concludes in a later study that the use of statins by children also does not result in loss of quality of life (Tonstad 2003). Likewise, De Jongh and colleagues observe no lower quality of life: they conclude that children feel safer with medication (De Jongh, Kerckhoffs et al. 2003). The conclusion that no large psychosocial problems are reported does not imply that the test and the treatment barely effect people's everyday lives or that it is easy to live with it. The absence of a negative effect on the quality of life does not imply by definition that there are no problems associated with the genetic testing of children: such reporting is the end result of the work that parents and children have to do in everyday life.

13   In the previous chapter it is described that much study of the clinical relevance of the many mutations involved in FH is still ongoing. Analogous to the question raised by Boenink in Chapter 3, we may also raise the normative question here whether knowledge of the clinical relevance of the various mutations is solid enough to make it *reasonable* to ask parents and children to do so much work. If some mutations present a strongly fluctuating clinical picture, should you ask parents and children to arrange their lives around FH? This is all the more relevant because we do not know the long-term effects of statins; the validity of the idea 'it can't do any harm and it may do some good' still needs to be established. Our findings do not solve this question. They give rise, however, to a careful consideration of the significance of providing more attention in the information given on FH to the possible effects of a positive result on the everyday life of those involved.

14   With parents who reject genetic testing of children this may be the case, but we do not know it because they were barely represented in our sample.

6   Genetics and insurance: new technologies, new policies, new responsibilites

1   For the sake of anonymity, the insurance companies are referred to here as Case 1 and Case 2. The fieldwork mainly consisted of observation of the underwriters' activities. From the beginning, I was assigned my own desk in the department which allowed me to keep track of those involved and their daily work. I began my fieldwork with introductory interviews with the management, medical advisers and some of the underwriters, so as to obtain a basic idea of the department and the general workings of the underwriting process. After some time I switched to observations of those involved. This required that I would sit next to individual underwriters to see how they actually *do* their work. This also involved looking at the 'devices' (Latour and Woolgar 1986) they use in making reference to insurance risks and that help them in coming to closure, such as medical questionnaires, handbooks, computers and guidelines. The materials gathered were analysed with the help of the software program Nvivo.

2   My use of 'trajectory' follows Strauss's definition of the term, highlighting the on-going, practice-based and thereby processual nature of ordering (Strauss 1993).

3　The moratorium and the WMK apply to life insurance and disability insurance, but not to health insurance (Breed Platform Verzekerden & Werk 2004).

4　It involves heredity testing performed on the person himself or his relatives. In life insurance the limit is fixed at an insured amount of 160,000 euro and in private disability insurance at an insured amount of 32,000 euro in the first year and 22,000 euro in subsequent years. Below this limit one does not have to indicate if a family member suffered or died from a serious, untreatable, hereditary disease (Dutch Association of Insurers 2003).

5　In this sense McGleenan comments on the Belgian law: 'An interesting question ... is whether it will impact on any predictive test which is indicative of an individual's "future state of health". If so, then routine prognostic procedures such as cholesterol and hypertension readings cannot be used for insurance purposes even when the release of such information would be to the advantage of the proposer' (McGleenan 2001, p. 61). Along the same lines, Chuffart responds: 'Strictly speaking, this article [Art. 95] implies that the applicant's cholesterol level or blood pressure cannot be used in the underwriting process as these measures are obviously predictive in nature!' (Chuffart 1997, p. 7).

6　This meshes with debates arising sociologically on the discourse of individualisation of health and the individual moral management of health risks (e.g., Baker and Simon 2002; Higgs 1998; Petersen and Bunton 1997). Within these perspectives, it is argued that the concept of health is nowadays more and more embroiled in that of citizenship. Health becomes associated with 'virtuousness' and managing one's relationship to health has become an important means by which individuals can express their ethical selves, fulfilling their obligations as good citizens.

7　Persons with a family history of Huntington's disease, however, were an exception.

8　This criticism regarding discrimination against persons with non-genetic disorders was also expressed in parliamentary debates on Articles 5 and 95 in the act's preliminary stage. One senator claimed, for instance, that high cholesterol had long been used to assess insurance risks and that hence there were no specific reasons to protect carriers of a genetic predisposition more than those with high cholesterol (Cousy and Schoorens 1994, p. 306).

9　Genetic risks, in other words, receive preferential treatment in comparison to other probabilistic information. In this respect, it is interesting to compare a positive HIV test with a genetic test (cf. Sandberg 1995). Until recently, the positive result of an HIV test was seen as a diagnosis of imminent and inevitable serious disease. These people's chance of survival has increasingly gone up over the last decade, however, in part on account of new experimental drugs. Studies also show that some HIV 'patients' will not even develop the disease. An HIV test, then, might well be less predictive than, for instance, a test for Huntington's disease.

10　Examples are the detection of phenylketonuria via blood tests or Tay-Sachs via measurement of mutated proteins.

11　Cf. Moore 2002. The biotech company Sciona, for instance, currently sells 'genetic lifestyle tests' whereby variations of nine genes can be detected (http://www.sciona.com/coresite/products.htm). This company claims that for £200 and based on these tests, it can offer individualized advice on your diet and lifestyle, 'to enable you to make informed decisions on how to improve your health and well-being'.

7   Work health and genetics: problems of regulation in a changing society

1   Suppose that of each 100,000 people, 100 are really carriers of a genetic trait (0.1 per cent). There is a test with a sensitivity (sensitivity to identify *all* real carriers) of 99 per cent and a specificity (sensitivity to select real carriers *exclusively*) of 99 per cent as well. The test thus identifies 99 of the 100 actual carriers as being a carrier. Of the 99,900 non-carriers, the test identifies 99/100 × 99,900 = 98,901 persons as non-carriers. 99,900 – 98,901 = 999 persons are thus mistakenly identified as being a carrier. The test identifies a total of 99 + 999 = 1098 persons as carriers and thus it suggests that the prevalence is over ten times higher than it actually is. Of those 1098 persons only 99/1098 × 100 = 9 per cent is really a carrier.

8   Genetic risks and justice in the workplace: the end of the protection paradigm?

1   As background information for this study, interviews were held with scientists, epidemiologists, company doctors and others involved in the field of work and health, in both the Netherlands and Denmark. This chapter, which makes use of the insights gained from the literature, is geared specifically to one aspect of the research involved, namely the relationships between protection, safety, responsibility and harm.

2   In addition, 'individual' factors play a role *prior to* deciding on a prevention policy, namely in fixing the limit values: the question is to what extent MAC values are based on the 'most sensitive employee'. I do not address further this concern in this chapter.

3   The solvent team report on the year 1997 claimed that among the 56 diagnosed cases of CTE there were 20 painters (36 per cent), 12 paint sprayers (21 per cent), 7 printers (13 per cent), and 6 home/project furnishers (11 per cent) (Ziekenfondsraad 1998)

4   The 1997 the SER proposed mandatory substitution for all non-stationary situations in which (high) exposure to organic solvents is either difficult or impossible to control. After discussing the policy with respect to OPS, the Dutch parliament came with a much more stringent policy that called for mandatory substitution in *all* situations in which the exposure is hard to control. It involved phased policy implementation.

5   In the report, the commission also claims: 'It is not possible to exclude differences in individual sensitivity to solvents. As long as individual exposure at work cannot be sufficiently quantified, it is impossible to establish if individual sensitivity is a major determining factor in the emergence of CTE' (Health Council of the Netherlands 1999, p. 38). For a study of the possible genetic aspects of individual sensitivity, see, for instance, Wenker (2001).

6   Individual factors mainly involve complaints from employers about employees who consider the protective gear supplied – such as surgical masks to filter

dust or smoke, ear protection against noise, and hard hats, clothing or shoes that offer physical protection – a nuisance: 'Employees tend to experience protective gear as a nuisance. Unfortunately its significance is only acknowledged after something goes wrong. The problem is that when it goes wrong, the employer is held accountable.' Interview with staff member ,Working Conditions FME-CWM, *Metalektro Profiel*, May 2003, pp. 10–11.

7   This involves both loss of income and damages. Recently, the willingness among employees to claim their loss of income has significantly gone up, in particular because of the fact that the material damage from occupational diseases in the Netherlands is no longer sufficiently covered collectively and compensated by disability benefits. Estimates of this 'claim market' vary from 700 million to 7 billion euro (Van Immerzeel 2004).

8   In the Dikmans/Unilever case the Dutch High Council judged as follows: 'This judgment ignores, after all, that when an employee has been exposed at work to substances that are hazardous to his health, a causal relationship – to be proved by the employee – has to be assumed if the employer refrained from taking the measures that in all due fairness are needed to prevent the employee from suffering harm in the performance of his work, and that, consequently, Unilever is already obliged now to indicate in detail whether it has indeed taken measures to this effect and if so, which ones' (quoted in Geers, Ruijgrok and van de Water 2003, pp. 1–13). Technically it meant the old Article 7A: 1638x, which stated that the employee must prove both the causal relationship and the employer's non-compliance with the duty to provide for health and safety in the workplace. Under the regime of the new Article 7:658, the burden of proof as to compliance rests with the employer. Based on this Article, the employee still has to prove in principle that the harm surfaced in the performance of work for his (former) employer. But in the evolving jurisprudence, however, the 'reversal rule' took effect: 'If a safety standard has been violated while the harm that has arisen should have been prevented precisely by adherence to this standard, the causal relation between the standard's violation and the harm inflicted is assumed, unless the offender proves that this harm would also have emerged without violation of the safety standard' (Geers, Ruijgrok and van de Water 2003, p. 6).

9   Apart from employers and employees, other actors are assumed as well; for example, the Employees Council, the Labour Inspectorate and the Arbo Agency (See Geers, Ruijgrok and van de Water 2003).

Bibliography

Alper, J. S. and J. Beckwith (1998), 'Distinguishing genetic from non-genetic medical tests: Some implications for anti-discrimination legislations', *Science & Engineering Ethics*, 4: 141–50.

American Journal of Medical Genetics (2003), Special issue: *Genetic Testing and the Family*, 119C(1).

Armstrong, D., S. Michie et al. (1998), 'Revealed identity: A study of the process of genetic counselling', *Social Science & Medicine*, 47(11): 1653–8.

Aronowitz, R. A. (1998), *Making Sense of Illness: Science, society, and disease* (Cambridge: Cambridge University Press).

Baker, T. and J. Simon (2002), 'Embracing risk. The changing culture of insurance and responsibility'. In: T. Baker and J. Simon (eds), *Embracing Risk: The changing culture of insurance and responsibility* (Chicago and London: The University of Chicago Press): 1–25.

Bakker H. D., A.Wiegman, J. C. Defesche and J. J. Kastelein (1997), 'Is opsporing en behandeling van familiare hypercholesterolemie geïndiceerd bij kinderen?', *Nederlands Tijdschrift voor Geneeskunde*, 141(52): 2548–51.

Barendrecht, J. M. (2004), 'Pak klassieke kern van rechtsstaat aan – regelgeving, geschiloplossing en democratie kunnen moderner worden georganiseerd', *de Volkskrant*, 1 July.

Bartley, M. (1985), 'Coronary heart disease and the public health', *Sociology of Health and Illness*, 7(3): 289–313.

Beck, U. (1992), *Risk Society – Towards a New Modernity* (London: SAGE Publications)

Benschop, R., K. Horstman and R. Vos (2003), 'Voice beyond choice: Hesitant voice in public debates about genetics in health care', *Health Care Analysis*, 11(2): 141–50.

Biesecker, B. B. (2000), 'Reproduction ethics. The ethics of reproductive counselling: nondirectiveness.' In: T. Murray and M. Mehlman (eds), *Encyclopedia of Ethical, Legal and Policy Issues in Biotechnology* (New York: John Wiley & Sons): 977–83.

Bijker, W. E. (1995), *Of Bicycles, Bakelites, and Bulbs – Toward a theory of sociotechnical change* (Cambridge, MA: MIT Press).

Bijker, W. E. and J. Law (eds), (1992), *Shaping Technology/Building Society: Studies in sociotechnical change* (Cambridge, MA: MIT Press).

Binsbergen, J. J., A. Brouwer, B. B. van Drenth, A. F. M. Haverkort and T. van der Weijden (1991), 'NHG-standaard cholesterol', *Huisarts en Wetenschap*, 34(12): 551–7.

Bloedlink (1998), *Wie de jeugd heeft, heeft de toekomst. Ondernemingsplan* (No Place: Bloedlink).

Bloedlink (1999) *Proactief.* (Newsletter), no. 1.

Bolt, I. (1997), *Goede raad. Over autonomie en goede redenen in de praktijk van de genetische counseling* (Amsterdam: Thesis Publishers).

Boltanski, L. (1999), *Distant Suffering: Morality, media and politics* (Cambridge: Cambridge University Press).

Borst, E. and H. Hoogervorst (2001), Letter to Parliament, 18 December 2001. Tweede Kamer Stuk nr. 28 127.

Bosk, C. (1993), 'The workplace ideology of genetic counselors.' In: D. M. Bartels, B. S. Leroy and A. L. Caplan (eds), *Prescribing Our Future* (New York: Aldine de Gruyter): 25–37.

Breed Platform Verzekerden en Werk (BPV&W), *Erfelijkheidsonderzoek en verzekeringen*, Web Page available at: http://www.bpv.nl/v5b.html [accessed 07/2004].

Breheny, N., E. Geelhoed, J. Goldblatt and P. O'Leary (2005), 'Cost-effectiveness of predictive genetic tests for familial breast and ovarian cancer', *Genomics, Society and Policy*, 1(2): 67–79.

Brinkgreve C. and B. van Stolk (1997), *Van huis uit. Wat ouders aan hun kinderen meegeven* (Amsterdam: Meulenhoff).

Brockett, P. L., R. MacMinn and M. Carter (1999), 'Genetic Testing, Insurance Economics and Societal Responsibility', *North American Actuarial Journal*, 3(1): 1–20.

Brugada, R. (2000), 'Role of molecular biology in identifying individuals at risk for sudden cardiac death', *The American Journal of Cardiology*, 86(9:I): K28–K33.

Buchanan, A. and D. W. Brock et al. (2000), *From Chance to Choice - Genetics and Justice* (Cambridge: Cambridge University Press).

Burke, W. (2002), 'Genetic testing', *New England Journal of Medicine*, 347(23): 1867–75.

Bus, J. (2000) 'Arbo-convenanten nieuwe stijl. Gezond werken is een lekkere worst', *Maandblad voor Arbeidsomstandigheden*, 76: 9.

Carmelli, D., A. C. Heath and D. Robinette (1993), 'Genetic analysis of drinking behavior in World War II Veteran Twins', *Genetic Epidemiology*, 10: 201–13.

Chuffart, A. (1997), *Genetics and Life Insurance: A few thoughts* (Zürich: Swiss Re.).

Clarke, A. (1991), 'Is non-directive genetic counselling possible?', *The Lancet*, 338: 998–1001.

Clarke, A (ed.), (1998), *The Genetic Testing of Children*. (Washington DC: Bios Scientific).

Cousy, H. and G. Schoorens (1994), *De Nieuwe Wet op de Landverzekeringsovereenkomst. Parlementaire voorbereiding van de Wet van 25 juni 1992 en van de wijzigende Wet van 16 maart 1994* (Antwerp: Kluwer Rechtswetenschappen).

Damme, van K. (2000), 'Genetic testing in the workplace: The scientific aspects.' In: *Genetic Testing in the Workplace – Proceedings of the Round Table Debate held at the Borchette Center*, Brussels, 6 March 2000 (Luxembourg: Office for the Official Publications of the European Communities. *Available at* http://europa.eu.int).

Davies, D. (1998), 'Health and the discourse of weight control.' In: A. Petersen and C. Waddell (eds), *Health Matters: A sociology of illness, prevention and care* (Buckingham: Open University Press): 141–55.

Davison, C. (1996), 'Predictive genetics: The cultural implications of supplying probable futures.' In: T. Marteau and M. Richards (eds), *The Troubled Helix: Social and psychological implications of the new human genetics* (Cambridge: Cambridge University Press): 317–30.

Daykin, C. D., D. A. Akers, A. S. MacDonald, T. Mcgleenan, D. Paul and P. J. Turvey (2003), Genetics and insurance – some social policy issues, *British Medical Journal*, 9: 787–874.

Department of Labor, Department of Health and Human Services, Equal Employment Opportunity Commission, Department of Justice (1998) *Genetic Information and the Workplace* (Washington DC: US Government).

Dewey, J. (1927), *The Public and Its Problems* (Athens, OH: Swallow Press / Ohio University Press).

Dicke, A. A. (1999), 'Discussion after Chambers 1999', *North American Actuarial Journal*, 3: 31–3.

Dijstelbloem, H., C. J. M. Schuyt and G. de Vries (2004), 'Dewey en de nieuwe politieke kwesties in de samenleving.' In: E. R. Engelen and M. Sie Dhian Ho (eds), *De Staat van de Democratie. Democratie voorbij de Staat* (Amsterdam: Amsterdam University Press).

Dreyfus, H. L. (in press), *Spinoza-lezingen UvA* (Assen: Koninklijke Van Gorcum) Dutch Cancer Society, see KWF.

Dutch Heart Foundation, see Nederlandse Hartstichting.

Dutch Association of Insurers (2003) Moratorium erfelijkheidsonderzoek Verbond van Verzekeraars. *Available at* http://www.verzekeraars.nl/download/Bijlage1D.pdf

Dykstra, A. P. (2003), *Het zit in de familie* (Utrecht: RU Utrecht).

EHC, *see* Stichting Erfelijke Hypercholesterolemie.

Ellsworth D. L., P. Sholinsky, C. Jaquish, R. R. Fabsitz and T. A. Manoloio (1999), 'Coronary heart disease: At the interface of molecular genetics and preventive medicine', *American Journal of Preventive Medicine*, 16(2): 122–33.

Elwyn, G., J. Gray and A. Clarke (2000), 'Shared decision-making and non-directiveness in genetic counseling', *Journal of Medical Genetics*, 37(2): 135–8.

Erkelens, D. W. (1993) *Nederlands Tijdschrift voor Geneeskunde*, 137, 28, 1420–1.

European Group on Ethics in Science and New Technologies to the European Commission (2000), *Genetic Testing in the Workplace* (Luxembourg: European Communities).

European Group on Ethics in Science and New Technologies to the European Commission (2003), *Ethical Aspects of Genetic Testing in the Workplace: Final Report* (Available at: http://europa.eu.int).

Evers, G. E. (2000), *Eigen baas zijn. Onderzoek naar zelfstandige ondernemers zonder personeel en hun bedrijf* (Hoofddorp: TNO Arbeid).

Ewald, F. (1999), 'Genetics, insurance and risk.' In: T. McGleenan, U. Wiesing and F. Ewald, *Genetics and Insurance* (Oxford: BIOS Scientific Publishers): 17–32.

Ewald, F. and J. P. Moreau (1994), 'Génétique médicale, Confidentialité et Assurance', *Risques*, 18: 111–30.

Fine, B. A. (1993), 'The evolution of nondirectiveness in genetic counseling and implications of the Human Genome Project.' In: D. M. Bartels, B. S. LeRoy and A. L. Caplan (eds), *Prescribing Our Future* (New York: Aldine de Gruyter): 101–17.

Finkler, K. (2000), *Experiencing the New Genetics: Family and kinship on the medical frontier* (Philadelphia: University of Pennsylvania Press).

FNV (2001), *Samenvatting toespraak bij de opening van bijeenkomst voor 'Vrienden'* Website Bureau Beroepsziekten FNV (Utrecht) (http://www.bbzfnv.nl).

FNV (2003), *Beschadigd bestaan. Een boekje open over beroepsziekten.* Website Bureau Beroepsziekten FNV (http://www.bbzfnv.nl).

FNV(2003a), *Rapport van Werkgroep 'Collectieve compensatie voor OPS slachtoffers'.* Website Bureau Beroepsziekten FNV (http://www.bbzfnv.nl).

FNV (2003b), 'Bouw Nieuws. FNV wil instituut voor hulp aan OPS-slachtoffers', *FNV Bouwmagazine*, 2(11): 14–16.

Fontaine, M. (1999), *Verzekeringsrecht* (Brussel: Larcier).

Foucault, M. (1994), *Dits et Écrits, 1954–1988* (Paris: NRF Édition Gallimard).

Freriks, D. (1994), 'De verzekeringswet van 25 juni 1992 en het medisch beroeps-geheim.' In: J. Van Steenberghe (ed.), *Medisch beroepsgeheim en verzekeringen* (Brugge: Die Keure): 13–34.

Fujimura, J. H. (1988), 'The molecular biological bandwagon in cancer research: Where social worlds meet', *Social Problems*, 35(3): 261–83.

Geers, A., J. Ruijgrok and R. van de Water (2003), 'RSI: de stand van zaken', *Nederlands Juristen Blad* (1–13 October).

Gevers Leuven, J. A. (1997), 'Opsporing familiaire hypercholesterolemie bij kinderen geïndiceerd? Bij uitzondering wel', *Nederlands Tijdschrift voor Geneeskunde* 141(52): 2551–4.

Gezondheidsraad (1989), *Erfelijkheid: Maatschappij en wetenschap – Over de mogelijkheden en grenzen van erfelijkheidsdiagnostiek en gentherapie* (Den Haag: Gezondheidsraad).

Gezondheidsraad (1999), *Piekblootstelling aan organische oplosmiddelen*. Publicatienr. 1999: 12 (Den Haag: Gezondheidsraad).

Gezondheidsraad (2001), *Prenatale Screening: Downsyndroom, neuralebuisdefecten, routine-echoscopie*. Publicatienr. 2001: 11 (Den Haag: Gezondheidsraad).

Gezondheidsraad (2004), *Prenatale Screening (2); Downsyndroom, neuralebuisde-fecten*. Publicatienr. 2004: 06 (Den Haag: Gezondheidsraad).

Gordijn, B. (ed.), (2004), *Spreken of zwijgen? Over de omgang met genetische tests* (Nijmegen: Valkhof Pers).

Graham, C. A., E. McClean, A. J. M. Ward, E. D. Beattie, S. Martin, M. O'Kane, I. S. Young and D. P. Nicholls (1999), 'Mutation screening and genotype: Phenotype correlation in familial hypercholesterolaemia', *Atherosclerosis*, 147: 309–16.

Grandstrand Gervais, K. (1993), 'Objectivity, value neutrality, and non-directiveness in genetic counseling.' In: D. M. Bartels, B. S. Lerou and A. L. Caplan, *Prescribing Our Future: Ethical challenges in genetic counselling* (New York: Aldine de Gruyter): 119–30.

Guldix, E., J. Stuy, K. Jacobs and A. Rigo (1994), *Het gebruik van genetische infor-matie. Het ethisch en juridisch kader voor het maatschappelijk gebruik van geïndivid-ualiseerde genetische informatie* (Brussels: Federale diensten voor wetenschappelijke, technische en culturele aangelegenheden).

Gunsteren, H. R. van and E. van Ruyven (eds), (1995), *Bestuur in de ongekende samenleving* (Den Haag: Sdu (Juridische and Fiscale Uitgeverij)).

Habermas, J. (2001), *Die Zukunft der menschlichen Natur. Auf dem Weg zu einer lib-eralen Eugenetik?* (Frankfurt a M: Suhrkamp Verlag).

Hageman, G., J. van der Hoek, M. van Hout et al. (1999), 'Parkinsonism, pyram-idal signs, polyneuropathy, and cognitive decline after long-term occupational solvent exposure', *Journal of Neurology*, 246: 198–206.

Have, A. M. J. ten (2001), 'Genetics and culture: The geneticization thesis', *Medicine, Health Care and Philosophy*, 4(3): 295–304.

Health Council of the Netherlands, see Gezondheidsraad.

Hedgecoe, A. M. (2001), 'Ethical boundary work: Geneticization, philosophy and the social sciences', *Medicine, Health Care and Philosophy*, 4: 305–9.

Hemerijck, A. (2002), 'Over institutionele aanpassing en sociaal leren – Een ver-handeling geïnspireerd door Albert Hirschmans trits van exit, voice and loyalty.' In: T. Jaspers and J. Outshoorn (eds), *De bindende kracht van concepten* (Amsterdam: Aksant): 5–48.

HGAC, see Human Genetics Advisory Commission.

Higgs, P. (1998), 'Risk, governmentality and the reconceptualization of citizenship.' In: G. Scambler and P. Higgs (eds), *Modernity, Medicine and Health: Medical sociology towards 2000* (London: Routledge): 177–97.

Hirschman, A. O. (1970), *Exit, Voice, and Loyalty – Responses to decline in firms, organizations, and states* (Cambridge, MA: Harvard University Press).

Hirschman, A. O. (1982), *Shifting Involvements – Private interest and public action.* (Princeton, NJ: Princeton University Press).

Hoeg, J. M. (1997), 'Evaluating coronary heart disease risk', *The Journal of the American Medical Association*, 277(17): 1387–90.

Hoek, J. A. F. van der (1998), 'Chronische toxische encefalopathie – De Solvent Team-benadering', *Tijdschrift Huisartsgeneeskunde*, 15(2): 77–83.

Horstman, K. (1997), 'Chemical analysis of urine for life insurance: The construction of reliability', *Science, Technology, & Human Values*, 22(1): 57–78.

Horstman, K. (2000), 'Preventie na 2000: tussen wetenschap en het goede leven', *Tijdschrift voor Gezondheidswetenschappen*, 2: 108–13.

Horstman, K. (2002), *Nooit meer ziek. Publieke en professionele verantwoordelijkheid voor bio-medical engineering* (Eindhoven: TU Eindhoven).

Horstman, K. (2004), Managers en uitvoerenden mogen geen neutrale vehikels zijn van overheidsbeleid', *Nederlands Tijdschrift voor Jeugdzorg*, 3: 142–9.

Horstman, K., G. H. de Vries and O. Haveman (1999), *Gezondheidspolitiek in een risicocultuur. Burgerschap in het tijdperk van de voorspellende geneeskunde* (Den Haag: Rathenau Instituut).

Hoyweghen, I. van (2004), *Making Risks: Travels in life insurance and genetics* (Leuven: Katholieke Universiteit Leuven, Faculteit Sociale Wetenschappen (proefschrift)).

Hoyweghen, I. van (2006), *Risks in the Making: Travels in life insurance and genetics.* (Amsterdam: Amsterdam University Press).

Hoyweghen, I. van, K. Horstman and R. Schepers (2006), 'Making the normal deviant: The introduction of predictive medicine in life insurance', *Social Science & Medicine*, 63: 1225–35.

Huibers, A. K. and A. van 't Spijker (1998), 'The autonomy paradox: Predictive genestic testing and autonomy: three essential problems', *Patient Education and Counselling*, 35: 53–62.

Human Genetics Advisory Commission (1999), *The Implications of Genetic Testing for Employment* (London: Human Genetics Advisory Commission).

Husted, J. (1997), 'Autonomy and a right not to know.' In: R. Chadwick, M. Levitt and D. Shickle (eds), *The Right to Know and the Right Not to Know* (Avebury: Aldershot): 55–68.

Illich, I. (1974), *Energy and Equity* (London, Calder & Boyars).

Immerzeel, M. van (2004), 'Schade door beroepsziekten onbetaalbaar', *de Volkskrant*, 11 May.

Jansen, N. W. H. (2003), *Working Time Arrangements, Work-Family Conflict, and Fatigue* (Maastricht: Universitaire Pers Maastricht).

Jasanoff, S. (2005), *Designs on Nature – Science and Democracy in Europe and the United States* (Princeton: Princeton University Press).

Jeger, N. and P. Cauwenbergh (1996/7), 'Individuele levensverzekeringen "overlijden" en erfelijkheidsonderzoek: een kritische analyse van de art. 5 en 95 van de Wet van 25 juni 1992 op de landverzekeringsovereenkomst', *Tijdschrift voor Gezozndheidsrecht/Revue de Droit de la Santé*: 239–56.

Jongh, S. de, M. C. Kerckhoffs, M. A. Groothuis, H. D. Bakker and B. F. Heymans Last (2002), *Familial Hypercholesterolemia in Childhood* (Amsterdam: UvA).

Jongh, S. de, M. C. Kerckhoffs, M. A. Groothuis, H. D. Bakker and B. F. Heymans Last (2003), 'Quality of life, anxiety and concerns among statin-treated children with familial hypercholesterolaemia and their parents', *Acta Paediatrica*, 92(9): 1096–101.

Jorde, L. B., J. C. Carey et al. (1999), *Medical Genetics* (St. Louis: Mosby).

Kastelein, J. J. P. and J. C. Defesche (1995), *Landelijke opsporing van patiënten met familiaire hypercholesterolemie door de Stichting Opsporing Erfelijke Hypercholesterolemie. Identificatie, diagnostiek en behandeling van personen met de heterozygote vorm van Familaire Hypercholesterolemie, met behulp van stamboomonderzoek en DNA-analyse.* Een beleidsplan voor een gerichte opsporing van personen met familiaire hypercholesterolemie (Amsterdam: StOEH).

Kelly, T. E. (1999), 'Genetic counselling after unexpected cytogenetic findings on prenatal diagnosis', *Southeast Asian Journal of Tropical Medical Public Health*, 30 (Suppl. 2): 183–5.

Keulartz, J., M. Korthals, M. Schermer and T. Swierstra (eds), (2002), *Pragmatist Ethics for a Technological Culture* (Deventer: Kluwer).

Khoury, M. J. (1997), 'Genetic epidemiology and the future of disease prevention and public health', *Epidemiologic Reviews*, 19(1): 175–81.

Khoury, M. J., W. Burke and E. J. Thomson (eds), (2000), *Genetics and Public Health in the 21st Century: Using genetic information to improve health and prevent disease* (Oxford: Oxford University Press).

Kirchner, D. B. (2002), 'The spectrum of allergic diseases in the chemical industry', *International Archives of Occupational and Environmental Health*, 75 (Suppl.): 107–12.

Kitcher, P. 1996, *The Lives to Come: The genetic revolution and human possibilities* (London: Penguin Books).

Knottnerus J. A. and E. Borst-Eilers (2002), 'Prenatale screening', *Medisch Contact*, 57(43): 1550–1.

KWF (2001), *Kanker in de familie. Hoe zit het met erfelijkheid?* (brochure).

KWF (2002), *Erfelijke borst – en eierstokkanker* (brochure).

Latour, B. (1987), *Science in Action: How to follow scientists and engineers through society* (Milton Keynes: Open University Press).

Latour, B. (1988), *The Pasteurization of France* (Cambridge, MA: Harvard University Press).

Latour, B. (1996), *ARAMIS* or the Love of Technology (Cambridge, MA: Harvard University Press).

Latour, B. and S. Woolgar (1986), *Laboratory Life: The Construction of Scientific Facts* (Princeton, NJ: Princeton University Press [1979]).

Latour, B. and P. Weibel (eds), (2005), *Making Things Public: Atmospheres of Democracy* (Cambridge, MA: MIT Press).

Lemmens, T. (2000), 'Selective justice, genetic discrimination, and insurance: Should we single out genes in our laws?', *McGill Law Journal*, 45: 347–412.

Lemmens, T. and L. Austin (2001), 'The challenges of regulating the use of genetic information', *Isuma. Canadian Journal of Policy Research*, 2(3): 26–37 (Available at: http://www.isuma.net [03/2003]).

Lente, H. van (1993), *Promising Technology: The dynamics of expectations in technological development* (Delft: Eburon).

Leschot, N. J. and H. G. Brunner (eds), (1998), *Klinische genetica in de praktijk* (Maarssen: Elsevier/Bunge).

Lippman, A. (1993), 'Prenatal genetic testing and geneticization: Mothers matter for all', *Fetal Diagnosis and Therapy*, 8(Suppl. 1): 175–88.

Loo, M. and A. Nieuwe Weme (1997), 'Organo Psycho Syndrome', *NVVK Info*, 6(3).

Lupton, D. (1993), 'Risk as moral danger: The social and political functions of risk discourse in public health', *International Journal of Health Services*, 23(3): 425–35.

Lupton, D. (1995), *The Imperative of Health: Public health and the regulated body* (London: Sage).

Marang-van de Mheen P. J., A. ten Asbroek, M. C. van Maarle, M. E. A. Stouthard, G. J. Bonsel and N. S. Klazinga (2000), *Screening op familiaire hypercholesterolemie in Nederland. Een evaluatie van kosten, effecten en maatschappelijke gevolgen.* (Amsterdam: AMC).

Marks, D., M. Thorogood, H. Andrew, W. Neil and S. E. Humphries (2003), 'A review on the diagnosis, natural history, and treatment of familair hypercholesterolaemia', *Atherosclerosis*, 168(1): 1–14.

McConkie-Rosell, A. and G. A. Spiridigliozzi (2004), '"Family matters": A conceptual framework for genetic testing in the family', *Journal of Genetic Counseling*, 13(1): 9–29.

McGleenan, T. (2001), *Insurance and Genetic Information* (London: Association of British Insurers (ABI)).

Meijers-Heijboer, E. J., L. C. Verhoog et al. (2000), 'Presymtomatic dna testing and prophylactic surgery in families with a BRCA1 or BRCA2 mutation', *The Lancet*, 355(9220): 2015–20.

Meijers-Heijboer, H., B. van Geel et al. (2001), 'Breast cancer after prophylactic bilateral mastectomy in women with a BRCA1 or BRCA2 mutation', *The New England Journal of Medicine*, 345(3): 159–64.

Mertens, F. J. H. (2003), 'Pech onderweg.' In: F. J. H. Mertens, R. Pieterman, C. J. M. Schuyt and G. de Vries, *Pech moet weg* (Amsterdam: Amsterdam University Press): 13–28.

Michie, S., F. Bron et al. (1997), 'Non-directiveness in genetic counseling: An empirical study', *American Journal of Human Genetics*, 60: 40–7.

Mills, C. W. (1959), *The Sociological Imagination* (New York: Oxford University Press).

Milunsky, A. (2001), *Your Genetic Destiny: Know your genes, secure your health, save your life* (Cambridge, MA: Perseus Publishing).

Mol, A. (1997), *Wat is kiezen? Een empirisch-filosofische verkenning* (Enschede: Universiteit Twente).

Mol, A. (2000), 'What diagnostic devices do: The case of blood sugar measurement', *Theoretical Medicine and Bioethics*, 21: 9–22.

Moore, P. (2002), 'Testing Times Ahead', *The Scientist*, 26 September. Web Page available at: http://www.biomedcentral.com/news/20020926/06/ (Accessed January 2004.)

Nederlands Huisartsen Genootschap, *see* NHG.

Nederlands Tijdschrift voor Geneeskunde, *see* NTvG.

Nederlandse Hartstichting (2001), *Erfelijkheid bij hart- en vaatziekten* (Den Haag: Nederlandse Hartstichting).

Nederlandse Organisatie voor Wetenschappelijk Onderzoek, *see* NWO.

Nelis, A. (1998), *DNA-diagnostiek in Nederland. Een regime-analyse van de ontwikkeling van de klinische genetica en DNA-diagnostische tests, 1970–1997* (Enschede: Twente University Press).

Nelis, A., G. de Vries and R. Hagendijk (2004), 'Stem geven' en "publiek maken" – wat patiëntenverenigingen ons kunnen leren over democratie', *Krisis. Tijdschrift voor empirische filosofie*, 5(3): 25–40.

Nelis, A., G. de Vries and R. Hagendijk (2007), 'Patients as Public in Ethics Debates – Interpreting the role of patient organisations in democracy.' In: P. Atkinson, P. Glasner and H. Greenslade (eds), *New Genetics, New Identities* (London: Routledge): 28 – 43.

NHG (Nederlands Huisartsen Genootschap) (1991), *Standaard Cholesterol* (Utrecht: NHG).

Niermeijer, M. F. et al. (2004), 'Meedelen en mededelen van erfelijkheidsgegevens in de familie: spreken of (ver)zwijgen door de "familieloze" mens. De geneticus als boodschapper?' In: B. Gordijn (ed.), *Spreken of zwijgen? Over de omgang met genetische tests* (Nijmegen: Valkhof Pers).

Nora, J. D., K. Berg and A. H. Nora (1991), *Cardiovascular Diseases: Genetics, epidemiology, and prevention* (New York and Oxford: Oxford University Press).

NWO (Nederlandse Organisatie voor Wetenschappelijk Onderzoek), (2003), *De bindende kracht van familierelaties* (Den Haag: NWO).

Nys, H. (1992), 'Van afkomst naar toekomst? Juridische grenzen van erfelijkheidsonderzoek bij verzekeringen', *Tijdschrift voor Verzekeringen/Bulletin des Assurances*, 299: 209–19.

Office of Technology Assessment, see OTA.

Ogilvie, C. M. (2003), 'Prenatal diagnosis for chromosomal abnormalities: Past, present and future', *Pathologie Biologie*, 51(3): 156–60.

Omenn, G.S. (2000) Public health genetics: an emerging interdisciplinary field for the post genomic era, *Annual Review Public Health*, 21: 1–13.

OTA (Office of Technology Assessment), (1983), *The Role of Genetic Testing in the Prevention of Occupational Diseases* (Washington DC: US Government Printing Office).

OTA (Office of Technology Assessment), (1990), *Genetic Monitoring and Screening in the Workplace* (Washington DC: US Government Printing Office).

Parker, Michael (2001), 'Genetics and the interpersonal elaboration of ethics', *Theoretical Medicine*, 22: 451–9.

Petersen, A. and R. Bunton (1997), *Foucault, Health and Medicine* (London: Routledge).

Petersen, A. and D. Lupton (1996), *The New Public Health: Health and self in the age of risk* (London: Sage).

Potting, M. (2001), *Van je familie. Zorg, familie en sekse in de mantelzorg* (Amsterdam: Aksant).

Press, N. (2000), 'Assessing the expressive character of prenatal testing: The choices made or the choices made available?' In: E. Parens and A. Asch (eds), *Prenatal Testing and Disability Rights* (Washington: Georgetown University Press): 214–33.

Proactief (1999/1), *Nieuwsbrief voor iedereen die te maken heeft met erfelijke hart- en vaatziekten* (Bennebroek: Stichting Bloedlink).

Raad voor Maatschappelijke Ontwikkeling, see RMO; see Tijmstra, T.

Rapp, R. (1999), *Testing Women, Testing the Fetus: The social impact of amniocentesis in America* (New York and London: Routledge).

Rapport Ziekenfondsraad (1998), *Solvent Team Project 1997 – Subsidiëring diagnostiek beroepsziekte OPS* (Amstelveen: Ziekenfondsraad, publ.nr. 793).

Rentmeester, C. A. (2001), 'Value neutrality in genetic counselling: An unattained ideal', *Medicine, Health Care and Philosophy*, 41: 47–51.

Riska, E. (2000), 'The rise and fall of Type A man', *Social Science & Medicine*, 51: 1665–74.

RMO (Raad voor Maatschappelijke Ontwikkeling), (2002), *Geen woorden maar daden – Bijdrage aan het Normen en waardendebat* (Den Haag: Sdu Uitgevers).

Roberts, R. (2000), 'A perspective: The new millennium dawns on a new paradigm for cardiology – molecular genetics', *Journal of the American College of Cardiology*, 36(3): 661–7.

Rose, N. (2002), *Lecture Presented at the Postgraduate Forum on Genetics and Society*, 10 September 2002 (Cambridge: Sidney Sussex College).

Rouvroy, A. (2000), 'Informations génétiques et assurance. Discussion critique autour de la position "prohibitioniste" du législateur belge', *Journal Des Tribunaux*, 119(5978): 585–602.

Sandberg, P. (1995), 'Genetic Information and life insurance: A proposal for an ethical European policy', *Social Science & Medicine*, 40(11): 1549–59.

Sarlo, K., and D. B. Kirchner (2002), 'Occupational asthma and allergy in the detergent industry: New developments', *Current Opinion in Allergy and Clinical Immunology*, 2: 97–101.

Schuermans, L. (1992/3), 'De nieuwe wet op de landverzekeringsovereenkomst', *Rechtskundig Weekblad*: 696.

Schuring-Blom, H. (2001), *Chromosomal Mosaicism in the Placenta: Presence and consequences* (thesis), (Amsterdam: University of Amsterdam).

Schuurman, W. M. M. (1995), *Blood Cholesterol: A public health perspective* (Wageningen: Wageningen University).

Schuyt, C. J. M. (1982), *Ongeregeld heden: naar een theorie van wetgeving in de verzorgingsstaat* (Alphen aan den Rijn: Samson).

Scott, S. and G. Williams (1991), 'Introduction.' In: S. Scott, G. Williams, S. Platt and H. Thomas (eds), *Private Risks and Public Dangers* (Avebury: Aldershot).

Senior, V., J. A. Smith, S. Mitchie and T. M. Marteau (2002), 'Making sense of risk: An interpretative phenomenological analysis of vulnerability to heart disease', *Journal of Health Psychology*, 7(2): 157–65.

SER Commissie Arbeidsomstandigheden (2002), *Normering piekblootstelling organische oplosmiddelen*. Publicatienummer 11 (Den Haag: SER).

SER Commissie Arbeidsomstandigheden (1997), *Preventie Organisch Psychosyndroom*. Publicatienummer 97/33 (Den Haag: SER).

Shklar, J. N. (1990), *The Faces of Injustice* (New Haven, CT, and London: Yale University Press).

Smith, R. (2002), 'In search of "non-disease"', *British Medical Journal*, 324(7342): 883–5.

Sobel, S. and D. Brookes Cowan (2000), 'The process of family reconstruction after DNA testing for Huntington Disease', *Journal of Genetic Counseling*, 9(3): 237–51.

Sociaal-economische Raad, *see* SER.

Steendam, G. van (1996), *Hoe genetica kan helpen. Een oefening in contextuele ethiek* (Leuven: Acco).

STG (Stichting Toekomstscenario's Gezondheidszorg), (2000), *Future of Predictive Medicine: A first step towards common policy* (Maarssen: Elsevier Gezondheidszorg).

Stichting Erfelijke Hypercholesterolemie, *Over leven met te hoog cholesterol in de familie* (brochure) (Leiden: Stichting EHC, no year).

Stichting voor Opsporing Erfelijke Tumoren, *see* StOET.

StOEH (Stichting Opsporing Erfelijke Hypercholesterolemie), (1995), *Jaarverslag* (Amsterdam: StOEH).

StOEH (1996), *Jaarverslag* (Amsterdam: StOEH).

StOEH (1997), *Jaarverslag* (Amsterdam: StOEH).

StOEH (1998), *Jaarverslag* (Amsterdam: StOEH).

StOEH (1999), *Aanvraag WBO vergunning* (Amsterdam: StOEH).

StOEH (1999a), Brochure (Amsterdam: StOEH).

StOEH (2000), Brochure (Amsterdam: StOEH).

StOET/Werkgroep Klinische Oncogenetica (2001), *Richtlijnenboekje voor diagnostiek en preventie erfelijke tumoren* (Amsterdam and Leiden: Stichting voor Opsporing Erfelijke Tumoren & Vereniging Klinische Genetica).

Straub, R. E., P. F. Sullivan, Y. Ma, M. V. Myakishev et al. (1999), 'Susceptibility genes for nicotine dependence: A genome scan and follow-up in an independent sample suggest that regions on chromosomes 2, 4, 10, 16, 17, 18 merit further study', *Molecular Psychiatry*, 4: 129–44.

Strauss, A. (1993), *Continual Permutations of Action* (New York: Aldine de Gruyter).

Sulston, J. and G. Ferry (2002), *The Common Thread – A Story of Science, Politics, Ethics and the Human Genome* (London: Bantam Press).

Swiss Re (2002). *Life underwriting experience study 2000: Mortality investigation based on data from the medical statistics of Swiss Re Zurich, 1965–1996* (Zürich: Swiss Re)

Temple, L. K. F., R. S. McLeod et al. (2001), 'Defining disease in the genomics era', *Science*, 293 (5531): 807–8.

Thomas S., T. van der Weijden, B. B. van Drenth, A. F. M. Haverkort, J. D. Hooi and J. D. van der Laan (1999), 'NHG-Standaard Cholesterol', *Huisarts en Wetenschap*, 42(9): 406–17.

Tijmstra, T. (2004), *Humane genetica en samenleving* (Den Haag: Raad voor Advies Maatschappelijke Ontwikkeling).

Tonkens, E. (2003), *Mondige burgers, getemde professionals. Marktwerking, vraagsturing en professionaliteit in de publieke sector* (Utrecht: NIZW).

Tonstad, S. (1996), 'Familial hypercholesterolaemia: A pilot study of parents' and children's concerns', *Acta Paediatrica*, 85(11): 1307–13.

Tonstad, S. (2003), 'Children and statins', *Acta Pediatrica*, 92(9): 1001–2.

Trip, M. D. (2002), *The Spectrum of Premature Atherosclerosis: From single gene to complex genetic disorder* (Amsterdam: Amsterdam University Press).

Trommel, W. and R. van der Veen (eds), (1999), *De herverdeelde samenleving: Ontwikkeling en herziening van de Nederlandse verzorgingsstaat* (Amsterdam: Amsterdam University Press).

Umans-Eckenhausen, M. A. W., J. C. Defesche, E. J. G. Sijbrands and R. L. J. M. Scheerder (2000), 'Review of the first 5 years of screening for familial hypercholesterolaemia in the Netherlands', *The Lancet*, 357: 165–8.

Verbond van Verzekeraars (2003), *Moratorium Erfelijkheidsonderzoek Verbond van Verzekeraars*. Available at: http://www.verzekeraars.nl/download/Bijlage1D.pdf

Voogd, A. C., E. J. Th. Rutgers and F. E. van Leeuwen (2002), 'Borstkanker – Beïnvloeden ontwikkelingen in de diagnostiek en behandeling de trends?', *Nationaal Kompas Volksgezondheid* (versie), 2(7), (Bilthoven: RIVM).

Vries, G. de (2003), 'Wat te doen met risico's?' In: F. J. H. Mertens, R. Pieterman, C. J. M. Schuyt and G. de Vries, *Pech moet weg* (Amsterdam: Amsterdam University Press): 32–49.

Vries, G. de (2005), 'Genetic screening at work: Risk and responsibility in the era of predictive medicine.' In: S.O. Hansson and E. Palm (eds), *The Ethics of Workplace Privacy* (Brussels: P.I.E.-PETER LANG S.A.): 17–38.

Weijden, van der T. (1993) *Nederlands Tijdschrift voor Geneeskunde*, 137, 28, 1420.

Wenker, M. (2001), *Individual Variation in Biotransformation: Relation to styrene kinetics and solvent-induced neurotoxicity* (dissertation), (Amsterdam: University of Amsterdam).

Werkgroep Collectieve Compensatie voor OPS Slachtoffers (2003), *Rapport Oktober 2003*, website Bureau Beroepsziekten FNV. Available at: http://www.bbbzfnv.nl

Wet op de Medische Keuringen (WMK), (1997), *Staatsblad* 365.

Wet van 25 juni op de Landverzekeringsovereenkomst (LVO), (1992), *Belgisch Staatsblad*, 20 August: 18283–333.

WHO (1998), *Report of a WHO Consultation; Paris, 3 October 1997: Familial Hypercholesterolemia* (Geneva: WHO).

Wiegman, A., J. Rodenburg, S. de Jongh, J. C. Defesche, H. D. Bakker, J. J. P. Kastelein and E. J. G. Sijbrands (2003), 'Family history and cardiovascular in familial hypercholesterolemia: Data in more than 1000 children', *Circulation*, 107: 1473–8.

Wilde, R. de (2000), *De voorspellers – een kritiek op de toekomstindustrie* (Amsterdam: De Balie).

Williams, C., P. Alderson et al. (2002), 'Is nondirectiveness possible within the context of antenatal screening and testing?', *Social Science and Medicine*, 54: 339–47.

Williams C, P. Alderson and B. Farsides (2002), 'Too many choices? Hospital and community staff reflect on the future of prenatal screening', *Social Science & Medicine*, 55: 743–53

Working Group Collective Compensations for OPS Victims (2003). Report available at: http:// www.bbzfnv.nl

Ziekenfondsraad, *see* Rapport Ziekenfondsraad.

ZON (Zorg Onderzoek Nederland), (2000), 'Verslag van de workshop naar aanleiding van het onderzoeksrapport', *Screening op familiare hypercholesterolemie in Nederland. Een evaluatie van kosten, effecten en maatschappelijke gevolgen* (Den Haag: ZON).

Zuuren, F. J. van (1997), 'The standard of neutrality during genetic counselling: An empirical investigation', *Patient Education and Counseling*, 32: 69–79.

Zwieten, M. van, D. Willems, L. Litjens, H. Schuring-Blom and N. Leschot (2005), 'How unexpected are unexpected findings in prenatal cytogenetic diagnosis?', *European Journal of Obstetrics & Gynaecology and Reproductive Biology*, 120(1): 16–22.

Zwieten, Myra van, Dick Willems, Lia Knegt and Nico Leschot (2006), 'Communication with patients during the prenatal testing procedure: An explorative qualitative study', *Patient Education and Counseling*, 63(1–2): 161–8.

Author Index

215

Subject Index